Caml … onal

AS & A L … ics:

Probability & Statistics 1

Practice Book

CAMBRIDGE
UNIVERSITY PRESS

CAMBRIDGE
UNIVERSITY PRESS

University Printing House, Cambridge CB2 8BS, United Kingdom

One Liberty Plaza, 20th Floor, New York, NY 10006, USA

477 Williamstown Road, Port Melbourne, VIC 3207, Australia

314–321, 3rd Floor, Plot 3, Splendor Forum, Jasola District Centre, New Delhi – 110025, India

79 Anson Road, #06–04/06, Singapore 079906

Cambridge University Press is part of the University of Cambridge.

It furthers the University's mission by disseminating knowledge in the pursuit of education, learning and research at the highest international levels of excellence.

www.cambridge.org
Information on this title: www.cambridge.org/9781108444903

First published 2018

20 19 18 17 16 15 14 13 12 11 10 9 8 7 6 5 4 3

Printed in Malaysia by Vivar Printing

A catalogue record for this publication is available from the British Library

ISBN 978-1-108-44490-3 Paperback

This Practice Book has been compiled and authored by Dean Chalmers, using some questions from:

Cambridge International AS and A Level Mathematics Statistics 1 Coursebook (Revised edition) by Steve Dobbs, Jane Miller and Julian Gilbey, that was originally published in 2016.

Cover image: Danita Delimont/Getty Images

Contents

How to use this book

Throughout this book you will notice particular features that are designed to help your learning. This section provides a brief overview of these features.

- Display numerical data in stem-and-leaf diagrams, histograms and cumulative frequency graphs.
- Interpret statistical data presented in various forms.
- Select an appropriate method for displaying data.

Learning objectives indicate the important concepts within each chapter and help you to navigate through the practice book.

TIP

On the evening when 30 people viewed films on screen A, there could have been as few as 37 or as many as 79 people viewing films on screen B.

Tip boxes contain helpful guidance about calculating or checking your answers.

WORKED EXAMPLE 2.1

The mass, x kg, of the contents of 250 bags of bird seed are recorded in the following table.

Mass (x kg)	$2.48 \leqslant x < 2.49$	$2.49 \leqslant x < 2.51$	$2.51 \leqslant x < 2.56$	$2.56 \leqslant x < u$
No. bags (f)	19	48	98	85

Given that the modal class is $2.49 \leqslant x < 2.51$, find to 2 decimal places the least possible value of u.

Answer

$\frac{85}{u - 2.56} < 240$ — For the modal class, frequency density $= \frac{48}{0.2} = 240$.

$85 < 240(u - 2.56)$

$u > 2.914$ — For $2.56 \leqslant x < u$, frequency density $= \frac{85}{u - 2.56}$

The least possible value of u is 2.92

Worked examples provide step-by-step approaches to answering questions. The left side shows a fully worked solution, while the right side contains a commentary explaining each step in the working.

E

Extension material goes beyond the syllabus. It is highlighted by a red line to the left of the text.

END-OF-CHAPTER REVIEW EXERCISE 1

1 A set of electronic weighing scales gives masses in grams correct to 3 decimal places.

Jan has recorded the masses, m grams, of a large number of small objects, and he finds that $5.020 \leqslant m < 5.080$ for 80 of them. Jan decides to illustrate the data for these 80 objects in a stem-and-leaf diagram.

a List an appropriate set of numbers that Jan can write into the stem of his diagram.

b Write down the least possible mass of any one of these 80 objects.

The **End-of-chapter review exercise** contains exam-style questions covering all topics in the chapter. You can use this to check your understanding of the topics you have covered.

Throughout each chapter there are exercises containing practice questions. The questions are coded:

PS These questions focus on problem solving.

P These questions focus on proofs.

M These questions focus on modelling.

You should not use a calculator for these questions.

You can use a calculator for these questions.

Chapter 1
Representation of data

- Display numerical data in stem-and-leaf diagrams, histograms and cumulative frequency graphs.
- Interpret statistical data presented in various forms.
- Select an appropriate method for displaying data.

1.1 Types of data and 1.2 Representation of discrete data: stem-and-leaf diagrams

WORKED EXAMPLE 1.1

Correct to the nearest centimetre, the length of each of the 80 pencils in a box is 18 cm.

a State the lower boundary and the upper boundary of the length of a pencil.

b What is the least possible total length of all the 80 pencils together?

c The 80 pencils are shared by six children so that each receives an odd number, and no two children receive the same number of pencils. Draw an ordered stem-and-leaf diagram showing one of the possibilities for the number of pencils given to the children.

Answer

a Lower boundary = 17.5 cm

Upper boundary = 18.5 cm

Lengths rounded to 18 cm mean $17.5 \leqslant \text{length} < 18.5$ cm.

b $80 \times 17.5 = 1400$ cm

c

Stem	Leaf
0	1 5 9
1	3
2	3 9

Key: 1 | 3 represents 13 pencils

The diagram must show six different, ordered odd numbers with a sum of 80. One possible solution, shown here, is to use 1, 5, 9, 13, 23 and 29.

EXERCISE 1A

1 Sara has collected three sets of data from the children in her daughter's class at school. These are: A: their first names; B: their heights; C: their shoe sizes. Match each set of data with the one word from the following list that best describes it.

X: discrete Y: qualitative Z: continuous

2 **a** Correct to the nearest ten metres, the perimeter of a rectangular football pitch is 260 metres. Complete the following inequality which shows the lower and upper boundaries of the actual perimeter:

.... m $\leqslant$ perimeter < .... m

b Eliana has 16 coins. The mass of each coin, correct to 1 decimal place, is 2.4 grams. Find the least possible total mass of the 16 coins.

3 A car was driven a distance of 364 km in five hours. The distance driven is correct to the nearest kilometre and the time taken is correct to the nearest hour.

Find the lower boundary and the upper boundary of the average speed of the car.

4 The numbers of items purchased by the first 11 customers at a shop this morning were 6, 2, 13, 5, 1, 7, 2, 11, 16, 20 and 15.

a Display these data in a stem-and-leaf diagram and include an appropriate key.

b Find the number of items purchased by the 12th customer, given that the first 12 customers at the shop purchased a total of 111 items.

5 **a** Correct to the nearest metre, the length and diagonal of a rectangular basketball court measure 18 m and 23 m, respectively. Calculate the lower boundary of the width of the court, correct to the nearest 10 cm.

b Correct to the nearest 0.01 cm, the radius of a circular coin is 0.94 cm. Find the least number of complete revolutions that the coin must be turned through, so that a point on its circumference travels a distance of at least 9.5 metres.

6 Bobby counts the number of grapes in 12 bunches that are for sale in a shop. To illustrate the data, he first produces the following diagram.

2	8 5 9
3	4 7 5 4
4	2 1 3 3 5

a State whether the data are:

i qualitative or quantitative

ii discrete or continuous.

b Complete Bobby's work by ordering the data in a stem-and-leaf diagram and adding a key.

7 Construct stem-and-leaf diagrams for the following data sets.

a The speeds, in kilometres per hour, of 20 cars, measured on a city street:

41, 15, 4, 27, 21, 32, 43, 37, 18, 25, 29, 34, 28, 30, 25, 52, 12, 36, 6, 25

b The times taken, in hours (to the nearest tenth), to carry out repairs to 17 pieces of machinery:

0.9, 1.0, 2.1, 4.2, 0.7, 1.1, 0.9, 1.8, 0.9, 1.2, 2.3, 1.6, 2.1, 0.3, 0.8, 2.7, 0.4

M 8 The contents of 30 medium-size packets of soap powder were weighed and the results, in kilograms correct to 4 significant figures, were as follows.

1.347	1.351	1.344	1.362	1.338	1.341	1.342	1.356	1.339	1.351
1.354	1.336	1.345	1.350	1.353	1.347	1.342	1.353	1.329	1.346
1.332	1.348	1.342	1.353	1.341	1.322	1.354	1.347	1.349	1.370

a Construct a stem-and-leaf diagram for the data.

b Why would there be no point in drawing a stem-and-leaf diagram for the data rounded to 3 significant figures?

PS 9 Two films are shown on screen A and screen B at a cinema each evening. The numbers of people viewing the films on 12 consecutive evenings are shown in the back-to-back stem-and-leaf diagram.

Screen A (12)		Screen B (12)
0	**3**	7
8 3	**4**	
7 6 4 0	**5**	3 4
7 4 1	**6**	4 5 6 7 8
9 2	**7**	1 6 8 9

Key: 1|**6**|4 represents 61 viewers for A and 64 viewers for B

A second stem-and-leaf diagram (with rows of the same width as the previous diagram) is drawn showing the total number of people viewing films at the cinema on each of these 12 evenings. Find the least and greatest possible number of rows that this second diagram could have.

TIP

On the evening when 30 people viewed films on screen A, there could have been as few as 37 or as many as 79 people viewing films on screen B.

10 The masses, to the nearest 0.1 g, of 30 Yellow-rumped and 30 Red-fronted Tinkerbirds were recorded by Biology students. Their results are given in the tables below:

Yellow-rumped					
17.0	13.1	15.7	14.7	17.4	15.3
14.2	16.2	16.9	16.8	16.5	14.2
15.5	16.9	13.5	18.1	17.6	17.9
13.0	15.1	18.2	17.8	18.1	12.3
12.2	17.7	16.5	16.7	14.8	16.6

Red-fronted					
18.2	14.7	15.9	14.7	15.2	14.5
17.3	13.2	14.0	20.2	17.5	15.6
19.3	19.4	17.4	16.5	16.8	15.8
15.3	13.3	16.0	18.6	13.6	18.0
14.1	15.5	20.0	13.9	16.8	14.7

a Display the masses in a back-to-back stem-and-leaf diagram.

b How many Yellow-rumped Tinkerbirds are heavier than the lightest 80% of the Red-fronted Tinkerbirds?

c The students decide to display the birds' masses, correct to the nearest 5 grams, in two bar charts.

   i Write down the frequencies for the three classes of Yellow-rumped Tinkerbirds.

   ii Explain why it is possible for the 20 g class of Red-fronted Tinkerbirds to have the same frequency as the 20 g class of Yellow-rumped Tinkerbirds.

   iii Given that there are equal numbers of Tinkerbirds in the two 20 g classes, construct two bar charts on the same axes showing the masses to the nearest 5 g.

1.3 Representation of continuous data: histograms

WORKED EXAMPLE 1.2

The percentage marks scored by 100 candidates in an examination are shown in equal-width intervals in the following histogram.

The marks are to be regrouped in four unequal-width intervals, as shown in the table.

Marks (%)	f	Frequency density
$9.5 \leqslant x < 19.5$	6	$6 \div 10 = 0.6$
$19.5 \leqslant x < 39.5$	a	
$39.5 \leqslant x < 69.5$	b	
$69.5 \leqslant x < 89.5$	c	

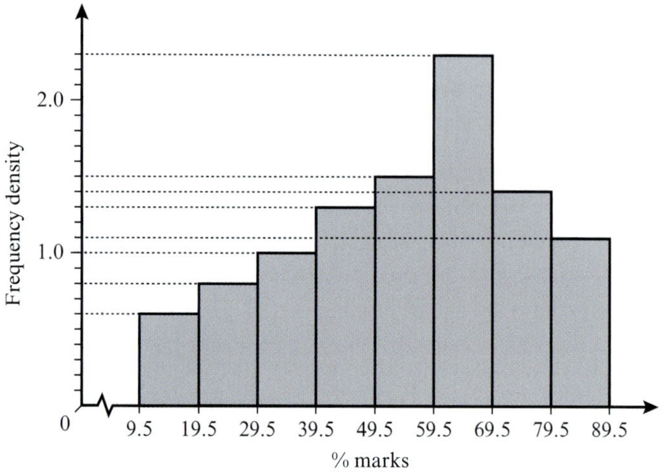

a Find the frequencies a, b and c.

b By calculating the three missing frequency densities, illustrate these data in a histogram with four unequal-width intervals.

Answer

a

Marks (%)	Column area	No. candidates (f)
$9.5 \leqslant x < 19.5$	$10 \times 0.6 = 6$	6
$19.5 \leqslant x < 29.5$	$10 \times 0.8 = 8$	8
$29.5 \leqslant x < 39.5$	$10 \times 1.0 = 10$	10
$39.5 \leqslant x < 49.5$	$10 \times 1.3 = 13$	13
$49.5 \leqslant x < 59.5$	$10 \times 1.5 = 15$	15
$59.5 \leqslant x < 69.5$	$10 \times 2.3 = 23$	23
$69.5 \leqslant x < 79.5$	$10 \times 1.4 = 14$	14
$79.5 \leqslant x < 89.5$	$10 \times 1.1 = 11$	11
	Total area = 100	$\Sigma f = 100$

Column area ∝ class frequency, and we know that the sum of the frequencies is 100. This allows us to draw up a grouped frequency table, which corresponds with the original histogram.

$a = 8 + 10 = 18$

$b = 13 + 15 + 23 = 51$

$c = 14 + 11 = 25$

Add together the relevant frequencies.

b For $19.5 \leqslant x < 39.5$, $fd = \dfrac{18}{39.5 - 19.5} = \dfrac{18}{20} = 0.9$

Frequency density (fd) $= \dfrac{\text{class frequency}}{\text{class width}}$

For $39.5 \leqslant x < 69.5$, $fd = \dfrac{51}{69.5 - 39.5} = \dfrac{51}{30} = 1.7$

For $69.5 \leqslant x < 89.5$, $fd = \dfrac{25}{89.5 - 69.5} = \dfrac{25}{20} = 1.25$

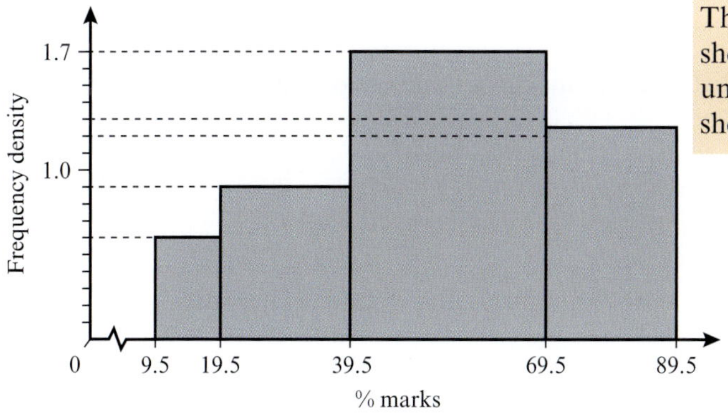

The required histogram showing the marks in four unequal-width intervals is shown.

EXERCISE 1B

1 The speeds, in km h^{-1}, of 200 vehicles travelling on a highway were measured by a radar device. The results are summarised in the following table. Draw a histogram to illustrate the data.

Speed	45 –	60 –	75 –	90 –	105 –	120 or more
Frequency	12	32	56	72	20	8

2 The mass of each of 60 pebbles collected from a beach was measured. The results, correct to the nearest gram, are summarised in the following table. Draw a histogram of the data.

Mass	5 – 9	10 – 14	15 – 19	20 – 24	25 – 29	30 – 34	35 – 44
Frequency	2	5	8	14	17	11	3

3 Thirty calls made by a telephone saleswoman were monitored. The lengths in minutes, to the nearest minute, are summarised in the following table.

Length of call	0 – 2	3 – 5	6 – 8	9 – 11	12 – 15
No. calls	17	6	4	2	1

a State the boundaries of the first two classes.

b Illustrate the data with a histogram.

4 The haemoglobin levels in the blood of 45 hospital patients were measured. The results, correct to 1 decimal place and ordered for convenience, are as follows.

9.1	10.1	10.7	10.7	10.9	11.3	11.3	11.4	11.4	11.4	11.6	11.8	12.0	12.1	12.3
12.4	12.7	12.9	13.1	13.2	13.4	13.5	13.5	13.6	13.7	13.8	13.8	14.0	14.2	14.2
14.2	14.6	14.6	14.8	14.8	15.0	15.0	15.0	15.1	15.4	15.6	15.7	16.2	16.3	16.9

a Form a grouped frequency table with eight classes.

b Draw a histogram of the data.

5 The table shows the age distribution, in whole numbers of years, of the 200 members of a chess club.

Age	16 – 19	20 – 29	30 – 39	40 – 49	50 – 59	over 59
No. members	12	40	44	47	32	25

a Form a table showing the class boundaries and frequency densities.

b Draw a histogram of the data.

6 The masses, w kilograms, of a selection of camera lenses are given in the following table.

Mass (w kg)	0.7–	1.2–	1.6–	1.9–	2.1–2.3
No. lenses (f)	24	56	75	36	9

a Give the masses of:

i the lightest 80 lenses

ii the heaviest 22.5% of the lenses.

b Illustrate the data in a fully labelled histogram.

c Estimate the number of lenses with masses of more than 2.0 kg.

7 The following table shows the lengths, to the nearest centimetre, of some chopsticks. Some of the class densities and some of the column heights that will be used to represent the data in a histogram are also shown.

Lengths (cm)	10–12	13–16	17–21	22–27
No. chopsticks (f)	12	24	15	c
Class density	4	6	b	2.5
Column height (cm)	8	a	6	d

Find the values of a, b, c and d.

PS 8 A histogram has been drawn with four columns that have unequal-width intervals. The areas of the four columns are, from left-to-right, 24, 44, 32 and 14 units^2.

a Find the total frequency of the data, given that the first column represents a frequency of 84.

b Explain why it would not be possible for the first column to represent a frequency of 54.

M 9 The blood glucose levels, in mmol/litre, of patients who attended a hospital during a particular week were recorded and are shown to the nearest tenth of a unit in the table.

Blood glucose (mmol/litre)	4.3–4.7	4.8–5.0	5.1–5.6	5.7–p	q–7.5
No. patients	35	27	66	45	r

a The data have a set of frequency densities that are symmetrical. Find the values of p, q and r.

b Of these patients, 10% are advised to take a new medication that may help to reduce their blood glucose levels. Without drawing a histogram, calculate an estimate of the range of these patients' current blood glucose levels, giving both boundary values correct to the nearest hundredth of a unit.

1.4 Representation of continuous data: cumulative frequency graphs

WORKED EXAMPLE 1.3

The distances, x km, covered by 250 delivery vehicles during the past week are summarised in the following cumulative frequency table.

Distance (x km)	<500	<800	<1200	<2000	<2400	<3000	<4000
No. vehicles (cf)	50	80	120	200	210	225	250

a What proportion of the vehicles covered distances from 1200 to 3000 km?

b Without drawing, show that a cumulative frequency polygon used to represent these data consists of just two straight-line segments.

Answer

a $\frac{105}{250} \times 100 = 42\%$

$225 - 120 = 105$ vehicles

b $(0, 0)$ to $(500, 50)$, $m = \frac{50 - 0}{500 - 0} = 0.1$

Investigate the gradient, m, between pairs of consecutively plotted points.

$(500, 50)$ to $(800, 80)$, $m = \frac{80 - 50}{800 - 500} = 0.1$

$(800, 80)$ to $(1200, 120)$, $m = \frac{120 - 80}{1200 - 800} = 0.1$

$(1200, 120)$ to $(2000, 200)$, $m = \frac{200 - 120}{2000 - 1200} = 0.1$

$(2000, 200)$ to $(2400, 210)$, $m = \frac{210 - 200}{2400 - 2000} = 0.025$

$(2400, 210)$ to $(3000, 225)$, $m = \frac{225 - 210}{3000 - 2400} = 0.025$

$(3000, 225)$ to $(4000, 250)$, $m = \frac{250 - 225}{4000 - 3000} = 0.025$

The polygon is made of one straight-line segment from $(0, 0)$ to $(2000, 200)$ and another from $(2000, 200)$ to $(4000, 250)$.

Write a brief conclusion to explain what the working shows about the polygon.

EXERCISE 1C

1 Estimates of the age distribution of a country for the year 2030 are given in the following table.

Age	under 16	16–39	40–64	65–79	80 and over
Percentage	14.3	33.1	35.3	11.9	5.4

a Draw a percentage cumulative frequency graph.

b It is expected that people who have reached the age of 60 will be drawing a state pension in 2030. If the projected population of the country is 42.5 million, estimate the number who will then be drawing this pension.

2 The records of the sales in a small grocery store for the 360 days that it opened during the year 2017 are summarised in the following table.

Sales, x (in \$100s)	$x < 2$	$2 \leqslant x < 3$	$3 \leqslant x < 4$	$4 \leqslant x < 5$
No. days	15	27	64	72
Sales, x (in \$100s)	$5 \leqslant x < 6$	$6 \leqslant x < 7$	$7 \leqslant x < 8$	$x \geqslant 8$
No. days	86	70	16	10

Days for which sales fall below \$325 are classified as 'poor' and those for which sales exceed \$775 are classified as 'good'. With the help of a cumulative frequency graph, estimate the number of poor days and the number of good days in 2017.

3 A company has 132 employees working in its city branch. The distances, x kilometres, that employees travel to work are summarised in the following grouped frequency table.

x (km)	<5	5–9	10–14	15–19	20–24	>24
Frequency	12	29	63	13	12	3

Draw a cumulative frequency graph. Use it to find the number of kilometres below which

a one-quarter of the employees travel to work

b three-quarters of the employees travel to work.

4 At a youth club, 80 girls and 80 boys were asked to estimate the duration of one minute by counting to 60 with their eyes closed. The results are shown in the following table.

Duration of count (seconds)	<50	<54	<58	<62	<66
No. girls (cf)	0	10	28	71	80
No. boys (cf)	0	32	52	63	80

a Draw and label a horizontal axis from 50 to 66 seconds using 1 cm for 1 second, and a vertical axis for cumulative frequency from 0 to 80 using 1 cm for 5 units.

b Plot five points for the girls and five points for the boys, then draw and label two cumulative frequency curves.

c Use your graphs to estimate the number of:

i girls who counted to 60 too quickly

ii boys who counted to 60 too slowly.

5 The cumulative frequency table shows the numbers of stamps, x, in 100 collections.

No. stamps (x)	$x < 40$	$x < 50$	$x < 80$	$x < 120$	$x < 150$	$x < 200$
No. collections (cf)	0	8	19	32	88	100

a How many of the collections contain 50 or more but fewer than 80 stamps?

b Find the least possible value of n, given that each of the largest 68 collections contains more than n stamps.

6 At a warehouse, cardboard is packed into bales for recycling. A full bale has a mass of 25 kg. The cumulative frequency table shows the number of full bales packed in the 52 weeks of last year. The data are to be displayed in a cumulative frequency polygon showing the mass of cardboard packed.

No. full bales	125 – 135	136 – 144	145 – 211	212 – 300
No. weeks (*cf*)	1	12	49	52

a Write down the coordinates of the two plotted points between which the polygon will be steepest.

b By drawing the polygon, or otherwise, estimate the number of weeks during which less than 4500 kg of cardboard was packed for recycling.

PS 7 The data displayed in the following stem-and-leaf diagram show the numbers of customers making purchases at two stores, A and B, over a period of 21 days.

Store A		Store B
9	**2**	5 5 5 6 7 8
3 2	**3**	0 1 3 3 4
7 7 6	**3**	5 5 8 9
4 4 3 1	**4**	0 1 1
9 9 7 7 5	**4**	6 8
4 4 3 2 0 0	**5**	0

Key: 9 | **2** | 5 represents 29 customers at A and 25 at B

a Show these data by drawing two cumulative frequency graphs on a single diagram.

On no two consecutive days during this period did the number of customers making purchases at store A decrease; on no two consecutive days during this period did the number of customers making purchases at store B increase.

b Draw up an ungrouped frequency table showing the numbers of customers who made purchases at both stores A and B on these 21 days.

P 8 a Under what conditions will a cumulative frequency graph representing the following set of data be a straight line?

Values of x	$x < x_1$	$x < x_2$	$x < x_3$	$x < x_4$	$x < x_5$
cf	0	a	b	c	d

b These data are represented by a straight-line cumulative frequency graph whose equation is $y = mx + c$. State the conditions under which the value of c is:

i negative

ii non-negative.

1.5 Comparing different data representations

WORKED EXAMPLE 1.4

The proportion of votes cast for the five election candidates, A, B, C, D and E, are represented in a pie chart. The chart has sector angles 25°, 86°, 92°, 58° and 99°, which are labelled A to E in that order.

a Explain why, from the information given, it is not possible to tell how many more votes candidate E received than candidate A.

b The winning candidate's son wishes to represent the votes cast for the candidates in a cumulative frequency graph. Give a reason why this type of diagram would not be appropriate in this case.

Answer

a We are not told the number of votes cast for any or all of the candidates.

b The data are not quantitative (and are not grouped).

EXERCISE 1D

1 Refer to the information about the votes cast for the five candidates in Worked example 1.4. If a total of 27 360 votes were cast, find the number of votes obtained by each candidate.

M 2 To illustrate her fellow students' favourite flavours of potato crisps, Mina draws a pie chart, labelling each sector with the flavour that it represents. She gives her diagram a suitable title and then writes the sector angles onto the chart.

a Give a reason why sector angles are not useful numbers to write onto the chart.

b Suggest two useful sets of numbers that she could use instead of sector angles.

M 3 The numbers of people employed at 60 companies are listed:

30	37	9	30	29	38	17	35	37	39	24	59	39	38	29
41	36	28	12	33	35	20	26	8	25	35	33	31	13	35
32	27	34	32	37	1	31	51	30	31	42	3	36	30	43
23	10	38	22	46	33	32	36	34	19	23	38	34	27	31

a How many of these companies employ more than 20 but not more than 35 people?

b Copy and complete the following grouped frequency table for the data.

No. employees	0–9	10–19	20–29	30–39	40–49	50–59
No. companies (f)	4					2

c A student is considering whether to illustrate the raw data in a stem-and-leaf diagram or to use the grouped frequency table to illustrate the data in a histogram.

Give one positive and one negative aspect of using each type of diagram in this case.

d The data are to be regrouped into three classes: 0 – 9, 10 – 39 and 40 – 59. A histogram will be drawn and the class 10 – 39 will be represented by a column of height 10 cm. Calculate the heights of the other two columns.

M 4 Last term, in a primary school of 275 children, 28% visited the library fewer than 10 times and $\frac{3}{25}$ visited more than 20 times.

a Briefly describe the group of children about whom no details are given in the information above.

b Illustrate the data about library visits in an appropriate and fully labelled diagram.

M 5 The scores of a cricketer in 40 consecutive innings are as follows:

6	18	27	19	57	12	28	38	45	66
72	85	25	84	43	31	63	0	26	17
14	75	86	37	20	42	8	42	0	33
21	11	36	11	29	34	55	62	16	82

a Illustrate the data on a stem-and-leaf diagram.

b State an advantage that the diagram has over the raw data. What information does not appear in the diagram that is given by the data?

6 At the start of a new school year, the heights of the 100 new students entering the school are measured. The results are summarised in the following table. '110 – ' means that those 10 students have heights not less than 110 cm but less than 120 cm.

Height (cm)	100 –	110 –	120 –	130 –	140 –	150 –	160 –
No. students	2	10	22	29	22	12	3

Use a graph to estimate the height of the tallest of the 18 shortest students.

M 7 A company employs 2410 people whose annual salaries are summarised as follows:

Salary (in $1000s)	< 10	10 –	20 –	30 –	40 –	50 –	60 –	80 –	⩾100
No. staff	16	31	502	642	875	283	45	12	4

a Draw a cumulative frequency graph for the grouped data.

b Estimate the percentage of staff with salaries between $26 000 and $52 000.

c If you were asked to draw a histogram of the data, what problem would arise and how would you overcome it?

8 The traffic noise levels on two city streets were measured one weekday, between 5.30 a.m. and 8.30 p.m. There were 92 measurements on each street, made at equal time intervals, and the results are summarised in the following grouped frequency table.

Noise level (dB)	<65	65–	67–	69–	71–	73–	75–	77–	⩾79
Street 1 frequency	4	11	18	23	16	9	5	4	2
Street 2 frequency	2	3	7	12	27	16	10	8	7

a Draw cumulative frequency graphs on the same axes for the two streets.

b Use them to estimate the highest noise level exceeded on 50 occasions, in each street.

c Write a brief comparison of the noise levels in the two streets.

M 9 The individual test marks (out of 25) of 150 students have been collected and are grouped in the following table.

Marks	1–5	6–10	11–15	16–20
No. students (f)	20	50	50	30

Two ways that could be used to represent these data are shown in the following diagrams.

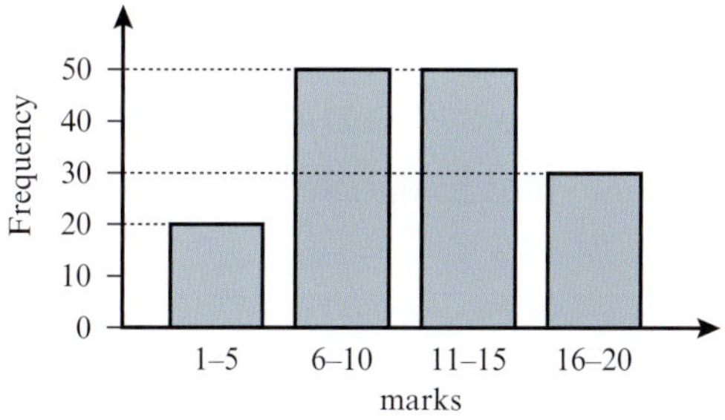

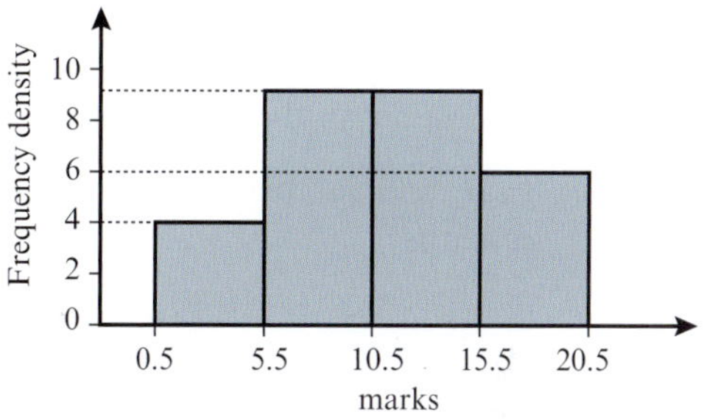

a Give some positive and some negative aspects of each method of representation.

b How would you represent the data if the marks were grouped in intervals of unequal widths?

END-OF-CHAPTER REVIEW EXERCISE 1

1 A set of electronic weighing scales gives masses in grams correct to 3 decimal places.

Jan has recorded the masses, m grams, of a large number of small objects, and he finds that $5.020 \leqslant m < 5.080$ for 80 of them. Jan decides to illustrate the data for these 80 objects in a stem-and-leaf diagram.

a List an appropriate set of numbers that Jan can write into the stem of his diagram.

b Write down the least possible mass of any one of these 80 objects.

2 A number of adults are classified by their age in whole years as 20 – 35, 36 – 55 or 56 – 60.

a In which class would Adam, who is 55 years and 11 months, be placed?

b State the mid-value of each of the three classes of ages.

3 Sixty vehicles are to be sold at auction. The following information about them is provided:

The distances that the vehicles have been driven (since new) range from 30 000 to 235 000 km.

There are 14 red, 11 black, 7 blue and 28 white vehicles.

All the vehicles have 2, 4, 6, 8 or 12 wheels.

The values of the vehicles are estimated to range from \$500 to \$135 000.

Copy and complete the table by placing ticks into the appropriate boxes to describe each set of data.

	Quantitative	Qualitative	Discrete	Continuous
Distance				
Colour				
No. wheels				
Value				

4 A financial advisor recorded the lengths of time that she spent in discussions with her clients during the last three months.

Length of time (minutes)	15 – 25	26 – 34	35 – 40	41 – 60
No. clients (f)	66	108	54	42

a Represent the data given in the table in a histogram.

b Estimate the number of clients with whom the advisor spent:

i less than 30 minutes

ii more than three-quarters of an hour.

5 Houses of various sizes have been built on a new estate. The numbers of bricks used to construct the external walls of these houses are given in the following cumulative frequency table.

No. bricks (x)	$x < 6300$	$x < 6500$	$x < 7000$	$x < 7500$	$x < 8000$	$x < 9000$
No. houses (cf)	0	14	32	56	68	75

a Draw up a grouped frequency table to illustrate the data about the numbers of bricks used.

b Represent the data from your grouped frequency table in a histogram.

c The largest ten of these houses each have four bedrooms. Estimate the least number of bricks that were used to construct the external walls of a four-bedroomed house.

6 A survey of the bank accounts of 2151 students at a university reveal that over 25% of them are in debt on the first day of term. Details about their balances are given in the following grouped frequency tables.

Debt (\$)	0–	500–	1200–3000
No. students (f)	300	280	36

Credit (\$)	0–	500–	1500–3000
No. students (f)	360	680	495

a Using this data, estimate the percentage of students with positive bank balances who have balances of over \$1000.

b Illustrate the data about student bank balances in a histogram.

c Find the correct frequency density that would be used if the debts of up to \$1200 were placed in a single class.

M 7 Eighty 100 gram samples of one brand of breakfast cereal were tested to assess how much protein (x) and dietary fibre (y) they contained.

Mass (grams)	6–	7–	10–	12–	15–18
No. samples (x)	2	6	22	44	6
No. samples (y)	28	36	8	4	4

a Determine how many of the samples contain:

i less than 10% protein by mass

ii between 10 and 15 grams of dietary fibre.

b Draw and label cumulative frequency graphs to illustrate these two sets of data.

c Based on evidence provided by these 80 samples, decide whether this brand of breakfast cereal contains more protein than dietary fibre. Give a reason for your answer.

8 The following table shows the number of people in the audience, to the nearest ten, at 240 evening performances of a play.

Number in audience	60 – 80	90 – 120	130 – 170	180 – 200
No. performances (f)	20	136	72	12

a Draw a cumulative frequency graph to illustrate this information.

b Estimate the number of performances at which there were between 100 and 150 people in the audience.

c The theatre is said to be 'running at a loss' when more than 60% of its 225 seats are unoccupied. Estimate the number of performances at which the theatre was running at a loss.

Chapter 2
Measures of central tendency

- Find and use different measures of central tendency.
- Calculate and use the mean of a set of data (including grouped data), either from the data itself or from a given total Σx or a coded total $\Sigma(x - b)$, and use such totals in solving problems which may involve up to two data sets.

2.1 The mode and the modal class

WORKED EXAMPLE 2.1

The mass, x kg, of the contents of 250 bags of bird seed are recorded in the following table.

Mass (x kg)	$2.48 \leqslant x < 2.49$	$2.49 \leqslant x < 2.51$	$2.51 \leqslant x < 2.56$	$2.56 \leqslant x < u$
No. bags (f)	19	48	98	85

Given that the modal class is $2.49 \leqslant x < 2.51$, find to 2 decimal places the least possible value of u.

Answer

$\frac{85}{u - 2.56} < 240$ — For the modal class, frequency density $= \frac{48}{0.2} = 240$.

$85 < 240(u - 2.56)$

$u > 2.914$ — For $2.56 \leqslant x < u$, frequency density $= \frac{85}{u - 2.56}$

The least possible value of u is 2.92

EXERCISE 2A

1 For the following distributions state, where possible, the mode or the modal class.

a

x	0	1	2	3	4
f	7	4	2	5	1

b

x	70	75	80	85	90
f	5	5	5	5	5

c

x	2–3	4–5	6–7	8–9	10–11
f	7	4	4	4	1

d

Eye colour	blue	brown	green
f	23	39	3

2 Find the mode(s) of the set of numbers 1.35, 0.64, 1.5, $\frac{1.28}{2}$, 2.25, 64%, $1\frac{1}{2}$, 0.8^2, 150%, $\sqrt{2.25}$ and $\sqrt{\frac{256}{625}}$.

M 3 In a quiz consisting of ten questions, the modal number of incorrect answers given by the five contestants was three. Interpret this information in terms of the number of correct answers given.

4 Three classes of continuous data, which are grouped as $5-8$, $9-13$ and $14-19$, have equal frequencies. Explain how you know that the modal class is $5-8$.

PS 5 Data about the heights of a group of adults are given in the following grouped frequency table. Find the greatest possible value of p, given that the modal class is not $1.68 \leqslant x < 1.72$ m.

Height (metres)	$1.65 \leqslant x < 1.68$	$1.68 \leqslant x < 1.72$	$1.72 \leqslant x < 1.80$	$1.80 \leqslant x < 1.96$
No. adults (f)	36	p	100	76

6 The total lengths of time that a group of 80 students spent on their homework assignments in one week are given in the table, to the nearest four hours. Find the mid-value of the modal class of times spent on homework assignments.

Time spent (hours)	0	4	8	12
No. students (f)	17	33	30	0

P 7 On a page of a particular book, the modal word, which occurs 28 times, is 'to'. Given that the page contains 487 words and that the word 'the' occurs 27 times, state the least and greatest possible number of times that the third most common word, 'love', appears.

PS 8 The table shows the frequency distribution of $7\frac{3}{4}k$ values of x, where k is a constant. Find the modal value of x.

x	k	$k+1$	$k+3$	$k+7$	$k+9$
Frequency	$k+11$	$k+13$	$2k$	$k-1$	$k-2$

2.2 The mean

Combined sets of data and means from grouped frequency tables

WORKED EXAMPLE 2.2

At a gymnastics event there are five male and seven female competitors. On the vault exercise, the males' mean score is 16.42 and the total score for the females is 121. Find the mean score on the vault for the male and female gymnasts together.

Answer

$$\frac{(16.42 \times 5) + 121}{5 + 7} = \frac{203.1}{12} = 16.925$$

Divide the total of the scores by the total number of gymnasts.

EXERCISE 2B

1 The numbers of aircraft that landed at an airfield on 30 consecutive days were recorded.

No. aircraft	5	6	7	8	9	10
No. days (f)	2	4	8	7	6	3

a Find the total number of aircraft that landed at the airfield during this period.

b Calculate the exact mean number of aircraft that landed per day.

2 The five numbers 69, 123, 234, 341 and 388 have a mean of 231. Determine by how much:

a the smallest of the five numbers must be increased so that the mean becomes 233

b the largest of the five numbers must be changed so that the mean becomes 211.

3 Four numbers are x, $3x$, $x - 24$ and $100 - x$.

a Find an expression in x for the mean of these numbers.

b Find the smallest and largest of these four numbers if their mean is equal to 42.

4 The mean of 25 numbers is equal to 16. The mean of the smallest 10 of these numbers is equal to 7. Find the mean of the largest 15 of these numbers.

5 The times taken in minutes to install an anti-virus program on 50 computers are summarised in the following grouped frequency table.

Time (t minutes)	$8 \leqslant t < 10$	$10 \leqslant t < 12$	$12 \leqslant t < 14$	$14 \leqslant t < 20$
No. computers (f)	13	22	10	5

a Write down the mid-value of each of the four classes of times.

b Calculate an estimate of the mean installation time, giving the exact answer in minutes.

c Give your estimate of the mean installation time correct to the nearest second.

6 The five numbers a, b, c, d and e have a mean of 137. Find the mean of the five numbers $100 - a$, $200 - b$, $300 - c$, $400 - d$ and $500 - e$.

7 A jeweller has in his stock 32 gold items with a mean mass of $53\frac{2}{5}$ grams and a number of silver items whose mean mass is $49\frac{2}{5}$ grams. The mean mass of all these items together is $51\frac{8}{11}$ grams. How many silver items does the jeweller have?

8 Boris has \$20 more than seven times the amount Ankit has.

Ankit has \$11 less than twice the amount Laila has.

In total, Boris and Ankit have \$8 more than 12 times the amount Laila has.

Find the mean amount of money that Ankit, Boris and Laila have.

9 Kitchen staff at a college prepare soup in 10 litre containers to serve to students at lunch time. The grouped frequency table shows the number of complete containers consumed each day last term.

No. complete containers	3 – 5	6 – 8	9 – 10	11 – 15
No. days (f)	2	5	49	14

Calculate an estimate of the mean quantity of soup consumed each day by the students, giving your answer to the nearest 100 millilitres.

M 10 a Find the mean of the two solutions to the equation $ax^2 + bx + c = 0$, where $a \neq 0$.

b Give a graphical interpretation of the value found in part **a**.

PS 11 Muneeza has 12 different-sized boxes that are all rectangular prisms. Each box has a capacity of 72 millilitres, and the sides of the boxes all measure an integer number of centimetres. One box, for example, measures 2 by 3 by 12 cm and another measures 1 by 8 by 9 cm.

Calculate the mean of:

a the surface areas of all the boxes

b the longest diagonals of all the boxes.

PS 12 The lengths, l cm, of 200 objects are summarised in the following grouped frequency table.

Length (l cm)	$9.0 \leqslant l < 10.2$	$10.2 \leqslant l < 11.4$	$11.4 \leqslant l < 15.0$	$15.0 \leqslant l < 16.4$
Frequency	30	44	56	70

a Calculate an estimate of the mean length.

b The boundary value of 11.4 cm is increased to 12.0 cm. A consequence of this is that the calculated estimate of the mean decreases. Find the least possible number of objects with lengths in the range $11.4 \leqslant l < 12.0$ cm.

PS 13 An ordinary fair die was rolled 100 times. The results were summarised in a frequency table and the mean score was calculated to be 3.46. It was later discovered that the frequencies, 15 and 20, of two consecutive scores had been swapped.

Find the least and greatest possible value of the true mean score.

Coded data

WORKED EXAMPLE 2.3

The number of occupants of a house is denoted by x. In a particular street of 50 houses, $\Sigma(x-2) = 106$. Find the mean number of occupants in these 50 houses.

Answer

$$\bar{x} = \frac{\Sigma(x-2)}{50} + 2$$

We use $\bar{x} = \frac{\Sigma(x-b)}{n} + b$.

$$= \frac{106}{50} + 2$$

$$= 4.12$$

Alternative method

$$\Sigma(x-2) = 106$$
$$\Sigma x - \Sigma 2 = 106$$
$$\Sigma x - 100 = 106$$
$$\Sigma x = 206$$

$$\bar{x} = \frac{206}{50} = 4.12$$

TIP

'Σ2' means the sum of the 50 '2s' that have been subtracted, i.e. 100.

EXERCISE 2C

1 Eighty values of x are such that $\bar{x} = 43.5$. Find the value of $\Sigma(x+1)$.

2 Forty values of y are summarised by the total $\Sigma(y-1) = 500$. Find $\bar{y}$.

P 3 Five boxes each contain some red sweets and some green sweets. The total number of sweets in a box is denoted by T. The number of red sweets and the number of green sweets in a box are denoted by R and G respectively.

In ascending order, the values of T are 11, 13, 17, 20 and 25.

In descending order, the values of $T - G$ are 21, 16, 13, 9 and 7.

a Find the mean number of red sweets in a box.

b What do the five boxes have in common?

4 The midday temperature, t°C, was recorded at a location on the Equator over ten days.

Given that $\Sigma(t-30) = 6$, find the mean temperature over the ten-day period.

5 For 40 values of x, it is given that $\Sigma(x+2) = 2 \times \Sigma(x-5)$. Find the mean value of x.

6 Twenty children, each carrying an identical heavy object, are weighed. The recorded masses of the children plus the object are denoted by m. Given that the object has a mass of 7 kg and that $\Sigma m = 1200$, find the mean mass of the children.

7 Ten girls were each given some money by their parents to spend at a fair. Each girl paid an entrance fee of \$2.50 and then they bought drinks that cost a total of \$18. Let the amount of money that each girl was given be x.

a Find an expression in terms of x for the total amount of money remaining after the entrance fee was paid, but before the drinks were bought.

b The mean amount remaining after they bought the drinks was \$8.20. What was the mean amount given to the girls by their parents?

8 Twenty values of x are such that $\Sigma(x - 1) = 586$. When these values of x are considered together with 30 values of y, the mean of all the values is 24.12. Find the value of $\Sigma(y + 1)$.

9 On a sports day, 20 adults and 30 children were timed running 200 metres. The times, in seconds, are denoted by a for the adults and by c for the children.

It is given that $\Sigma(a - 20) = 188$ and $\Sigma(c - 30) = 237$.

a Show that, on average, the children took 8.5 seconds longer than the adults to run 200 metres.

b Find the mean time taken by the adults and children together.

E

WORKED EXAMPLE 2.4

The cost of hiring a taxi is made up of a fixed charge of \$5 plus \$1.20 per kilometre travelled.

a Write down an expression for the cost in dollars of a taxi journey of x km.

b Arthur made six separate taxi journeys, for which he paid a total of \$210. Find the mean distance that he travelled on these journeys.

Answer

a $1.2x + 5$

b $\Sigma(1.2x + 5) = 210$

$1.2\Sigma x + \Sigma 5 = 210$

'Σ5' means the sum of the six fixed charges of \$5, i.e. \$30.

$1.2\Sigma x + 30 = 210$

$\Sigma x = \frac{(210 - 30)}{1.2}$

$\Sigma x = 150$

$\bar{x} = \frac{150}{6} = 25$ km

Alternative method

$\bar{x} = \frac{1}{1.2} \times \left(\frac{210}{6} - 5\right) = 25$ km

TIP

Alternatively, we can use: $\bar{x} = \frac{1}{a} \times [\text{mean of } (ax - b) + b]$ where a = 1.2 and b = −5.

EXERCISE 2D

1 Fifty data values denoted by x are such that $\Sigma x = 200$. Evaluate k such that $\Sigma(2x + k) = 0$.

2 A set of n data values is denoted by y and it is given that $\Sigma\left(\frac{1}{2}y - 1\right) = 56$ and that $\bar{y} = 18$. Find the value of n.

P 3 Prove that a density measured in kg m^{-3} is numerically identical to a density measured in grams per litre.

M 4 On a particular day the exchange rates involving three currencies were £1 to \$1.2603 and \$1 to €0.9082. The profits made on five short-term investments on that day are denoted by x, which is measured in euros (€).

a The mean profit made on these investments in pounds (£) was £4281.50. Find the value of Σx.

b What assumption are you making in your calculations?

5 The 12 values in each of two sets of related data are denoted by x and y.

a If $y = 3x - 2$, find $\bar{y}$, given that $\Sigma x = 84$.

b If $5x + 3y = 13$, find $\bar{x}$, given that $\Sigma y = -18$.

6 The times taken, in seconds, by ten athletes to complete an 800 metre race were recorded and are denoted by t. It is given that

$$\Sigma\left(\frac{t}{60} - 2\right) = \frac{27}{10}$$

Find, in minutes, the mean time taken to complete the race.

PS 7 Three of the vertices of rectangle P are at (1, 1), (–2, 1) and (1, 3).

P is mapped to Q by an enlargement through the origin with scale factor 2, followed by a reflection in the line $y - x = 0$.

a Find $(\bar{x}_Q, \bar{y}_Q)$, the coordinates of the centre of Q.

b For any rectangle whose centre is at $(\bar{x}, \bar{y})$, find an expression for the coordinates of the centre of the image after an enlargement through the origin with scale factor 2, followed by a reflection in the line $y - x = 0$.

PS 8 To furnish her new home, a woman purchased n items at a store. The mean cost of an item, inclusive of 10% sales tax and a delivery fee of \$5, is denoted by $\bar{x}$.

a Write down an expression in $\bar{x}$ for the mean cost of an item before the sales tax and delivery fee were included.

b Find the number of items that the woman bought, given that the mean amount of tax paid on an item was \$6.50 and that $\Sigma x = 2448$.

2.3 The median

Estimating the median and choosing an appropriate average

WORKED EXAMPLE 2.5

The times taken, to the nearest second, for 50 similar chemical reactions to take place are summarised in the following grouped frequency table.

Time (seconds)	20 – 29	30 – 39	40 – 49	50 – 54	55 – 70
No. reactions (f)	11	a	b	7	3

Given that the median time is in the class 30 – 39, find the least possible value of a and of b.

Answer

$11 + a \geqslant 25$

$a \geqslant 14.$

The least possible value of a is 14.

For 50 items of grouped data, the median is the value in the $\frac{50}{2} = 25$th position.

The least possible value of b is 0.

$\Sigma f = 11 + a + b + 7 + 3 = 50$, so $a + b = 29$.

The least possible value of b occurs when a takes its greatest possible value, which is 29.

EXERCISE 2E

1 Find the median mass of 6.6 kg, 3.2 kg, 4.8 kg, 7.6 kg, 5.4 kg, 7.1 kg, 2.0 kg, 6.3 kg and 4.3 kg.

A mass of 6.0 kg is added to the set. What is the median of the ten masses?

2 The number of rejected CDs produced each day by a machine was monitored for 100 days.

The results are summarised in the following table. Estimate the median number of rejects.

No. rejects	0 – 9	10 – 19	20 – 29	30 – 39	40 – 49	50 – 59
No. days	5	8	19	37	22	9

P 3 A list contains n numbers in ascending order, where n is odd.

a Write down, in their simplest form and in terms of n, the positions of the two values adjacent to the median.

b Using $n = 5$, give an example of a case where the mean of the two values adjacent to the median is:

i equal to the median

ii greater than the median

iii less than the median.

4 An arithmetic test of eight questions was given to a class of 32 students. The results are summarised in the following table.

No. correct answers	0	1	2	3	4	5	6	7	8
No. students	1	2	1	4	4	6	7	4	3

a Find the mean, median and mode of the number of correct answers. Interpret the median and mode in the context of this arithmetic test.

b Describe the shape of the distribution.

5 The ten numbers 5, 15, 12, 7, 34, 28, 17, 41, x and $x + 5$ have a mean of 18.

a Find the value of x.

b Find the median number.

c Explain why this set of numbers has no mode.

6 A list of numbers contains ten 5s, six 6s, nine 7s and eight 8s. Find the mode, the mean and the median.

P 7 The following diagram shows part of a cumulative frequency graph illustrating the values of x.

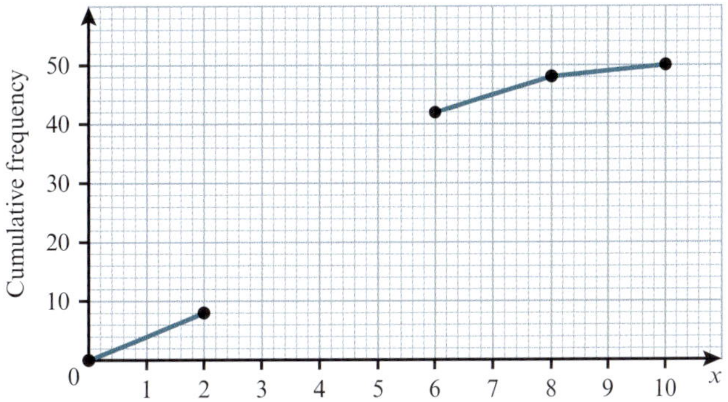

a Find the value of the cumulative frequencies b and c used in the following table.

x	$x < 0$	$x < 2$	$x < 4$	$x < 6$	$x < 8$	$x < 10$
cf	0	8	a	b	c	50

b Find the value of a in the previous table if the median of x is equal to:

i 3.4

ii 4.6.

c State the least and the greatest possible value of the median.

8 The masses, m kilograms, of the 400 babies born at a maternity hospital last year are summarised in the following cumulative frequency table.

Mass (m kg)	$m < 2.6$	$m < 3.0$	$m < 3.3$	$m < 3.6$	$m < 3.8$	$m < 4.1$
No. babies (cf)	0	195	295	345	385	400

An estimate for the median that would be obtained from a cumulative frequency polygon can be calculated as follows:

$$3.0 + \frac{200 - 195}{295 - 195} \times (3.3 - 3.0) = 3.015 \text{ kg}$$

Use a similar method to calculate an estimate of the interquartile range of the masses, giving your answer to an appropriate degree of accuracy.

9 The following table summarises the maximum daily temperatures in two holiday resorts in July and August 2017.

Temperature (°C)	18.0–19.9	20.0–21.9	22.0–23.9	24.0–25.9	26.0–27.9	28.0–29.9
Resort 1 frequency	9	13	18	10	7	5
Resort 2 frequency	6	21	23	8	3	1

a State the modal classes for the two resorts.

b A student analysed the data and came to the conclusion that, on average, Resort 1 was hotter than Resort 2 during July and August 2017. Is this conclusion supported by your answer to part **a**? If not, then obtain some evidence that does support the conclusion.

M 10 In some market research, 125 householders in various locations were asked to record how long it took to boil one litre of water using their electric kettles. The times are given to the nearest second:

Time (seconds)	180–190	191–199	200–210
No. householders (f)	3	90	32

a Estimate, to the nearest second, the median time taken to boil one litre of water.

b Show that a calculated estimate of the mean time taken is 197.32 seconds.

c It was later discovered that the three times in the class 180–190 were exactly 187, 188 and 189 seconds. Without further calculation, state how the estimates of the median and mean would be affected if this information were taken into account.

END-OF-CHAPTER REVIEW EXERCISE 2

1 The following table shows the number of students in the 40 classes at a secondary school.

No. students	24 – 28	29	30	32
No. classes (f)	5	15	12	8

a State the median and the mode of the data.

b Calculate the mean number of students in the largest 20 classes.

c Calculate an estimate of the mean number of students in the classes at this school.

2 The following tables show the frequencies of values of two variables x and y. For the data shown in the first table, it is given that $\bar{x} = 12.4$.

Calculate an estimate of the mean value of y.

x	10	16	22	28
Frequency	p	q	r	s

y	0 –	6 –	12 –	18 – 24
Frequency	p	q	r	s

3 In an experiment involving x and y, 40 readings of x are such that $\Sigma x = 12.905$ and 35 readings of y are such that $\bar{y} = -1.288$. Find the mean of the 75 readings of x and y considered together.

4 A cyclist travels 800 metres uphill for 5 minutes and then cycles downhill at an average speed of $32.4\ \text{km h}^{-1}$ for $4\frac{1}{2}$ minutes.

Calculate the cyclist's average speed in metres per second for the whole journey.

PS 5 A broadcasting company employs males and females in the ratio 3:2. The total of the annual salaries of the males and females are in the ratio 4:3. Find the ratio of the males' mean salary to the females' mean salary, giving your answer in its simplest form.

PS 6 The table gives the masses, to the nearest tonne, of some freight containers.

Mass (tonnes)	5 – 7	8 – 10	11 – 14	15 – 20	21 – 25
No. containers	21	q	27	50	31

a Given that the only modal class is 8 – 10 tonnes, find the least possible value of q.

b Find the greatest possible value of q, given that a calculated estimate of the mean is greater than 14.5 tonnes.

7 A computer has randomly selected n numbers which are denoted by x. Given that $\Sigma(x - 2.5) = -11.8$ and that $\Sigma(x + 1.5) = 24.2$, find the value of n.

P 8 The following frequency table shows 79 values of t.

t	16	17	18	19	20
Frequency	4	20	27	23	5

A reduction in the measures of central tendency for t is required. This is to be done by reducing some of the $t = 18$ values to 17.

Find the least number of $t = 18$ values that need to be reduced, so that the reduction occurs in the value of:

a the mode

b the median

c the mean.

9 Two airline companies, A and B, were asked to note how many passengers ordered vegetarian meals on international flights. Company A collected data on n flights and company B collected data on $n + 20$ flights.

The numbers of passengers who ordered vegetarian meals are denoted by a and by b for the two companies respectively, and are summarised by the totals $\Sigma(a - 6) = 18$ and $\Sigma(b - 7) = 18$.

Find the value of n, given that the mean number of passengers who ordered vegetarian meals on all the flights for which data were collected was 6.96.

Chapter 3
Measures of variation

- Find and use different measures of variation.
- Use a cumulative frequency graph to estimate medians, quartiles and percentiles.
- Calculate and use the standard deviation of a set of data (including grouped data) either from the data itself or from given totals Σx and Σx^2, or coded totals $\Sigma(x-b)$ and $\Sigma(x-b)^2$, and use such totals in solving problems which may involve up to two data sets.

3.1 The range and 3.2 The interquartile range and percentiles

Ungrouped data, grouped data and box-and-whisker diagrams

WORKED EXAMPLE 3.1

The cumulative frequency table summarises the heights, h cm, of 136 plants at a nursery.

Height (h cm)	$h < 4$	$h < 10$	$h < 16$	$h < 18$	$h < 20$	$h < 25$	$h < 30$	$h < 35$
No. plants (cf)	0	14	34	68	82	102	120	136

Represent the data in a box-and-whisker diagram.

Answer

$\Sigma f = 136$, so

Q_1 is the $\dfrac{136}{4} = 34$th value, which is 16 cm

Q_2 is the $\dfrac{136}{2} = 68$th value, which is 18 cm

Q_3 is the $\dfrac{3 \times 136}{4} = 102$nd value, which is 25 cm

The boundary values are 4 cm and 35 cm.

To create the diagram, we need to know the three quartiles and the boundary values of the data.

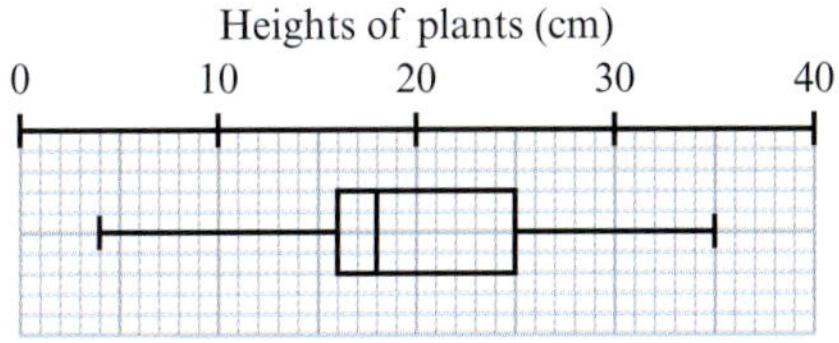

The diagram is shown with a scale.

EXERCISE 3A

1 Find the range and interquartile range of each of the following data sets.

a 7 4 14 9 12 2 19 6 15

b 7.6 4.8 1.2 6.9 4.8 7.2 8.1 10.3 4.8 6.7

2 The number of times each week that a factory machine broke down was noted over a period of 50 consecutive weeks. The results are given in the following table.

No. breakdowns	0	1	2	3	4	5	6
No. weeks	2	12	14	8	8	4	2

Find the interquartile range of the number of breakdowns in a week.

3 The seven numbers x, 11, 29, 56, 44, 21 and 32 have a range of 49. Find the two possible values for:

a x

b the interquartile range.

4 The lengths, x cm, of 99 objects are given in the grouped frequency table.

Length (x cm)	$0 \leqslant x < 8$	$8 \leqslant x < 16$	$16 \leqslant x < 24$	$24 \leqslant x < 30$
No. objects (f)	24	29	21	25

a Explain how you know that the lower quartile length is in the class $8 \leqslant x < 16$ cm.

b In which class is the upper quartile length?

c Find the least and greatest possible interquartile range of the lengths.

5 The audience size in a theatre was monitored over a period of one year. The sizes for Monday and Wednesday nights are summarised in the following table.

Audience size	50 – 99	100 – 199	200 – 299	300 – 399	400 – 499	500 – 599
No. Mondays	12	20	12	5	3	0
No. Wednesdays	2	3	20	18	5	4

Compare the audience sizes on Mondays and Wednesdays.

6 Values of u are summarised in the following box-and-whisker diagram.

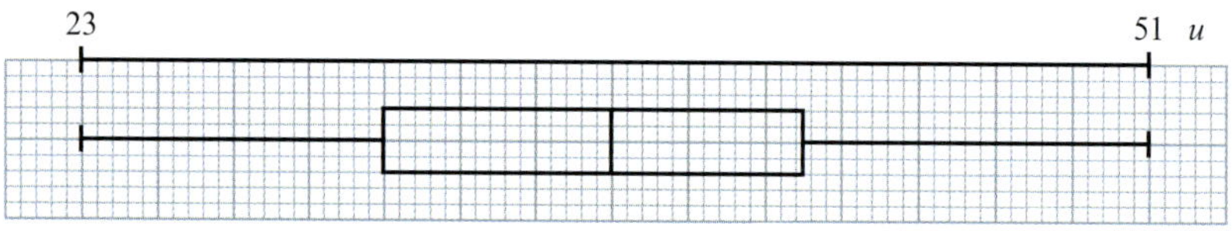

Find the range, the median and the interquartile range of these values of u.

7 Draw a box-and-whisker diagram to summarise the values of v given in the following table.

Range	Lower quartile	Median	Interquartile range	Maximum
26	59	65	13	76

8 The lengths, in cm, of 19 oak leaves are shown ordered as follows:

2.3, 2.6, 2.7, 2.8, 3.0, 3.1, 3.2, 3.5, 3.6, 3.8, 4.3, 4.4, 4.9, 4.9, 5.6, 5.9, 6.4, 6.8, 7.2

a Present these data in a simple stem-and-leaf diagram.

b Use your diagram to identify the median length and the interquartile range.

c Construct a box-and-whisker diagram of these data.

PS 9 A box-and-whisker diagram is drawn to summarise a set of data. The lowest value in the set of data is at distances of 6.2, 8.8, 10.7 and 13.5 cm from the other four values on the diagram. Find the interquartile range of the data, given that its range is 117 units.

PS 10 A cumulative frequency polygon representing values of x is drawn by plotting points at (0, 0), (1, a), (3, b), (7, c) and (10, d). The ratio of the gradients of the four line segments of the polygon are, in the same order as written previously, 4 : 3 : 2 : 1.

Given that the total frequency of the data is 105, estimate:

a the median value of x

b the range of the middle 60% of the values of x.

11 An examination paper has a number of questions all of equal difficulty. In a two-hour examination the number of questions answered by a random sample of 1099 candidates is shown in the table.

No. questions	0–4	5–9	10–14	15–19	20–24	25–29	30–34
No. candidates	12	98	308	411	217	50	3

a Construct a cumulative frequency table and draw a cumulative frequency graph.

b Use your graph to estimate the median and the interquartile range of these data.

c It should be possible for 3% of candidates to answer all questions on the exam paper. Find how many questions there should be on the two-hour paper.

12 The following box-and-whisker diagrams illustrate the scores in an aptitude test taken by people applying for a job. The scores are expressed on a scale of 0–50, and the results for men and women are shown separately.

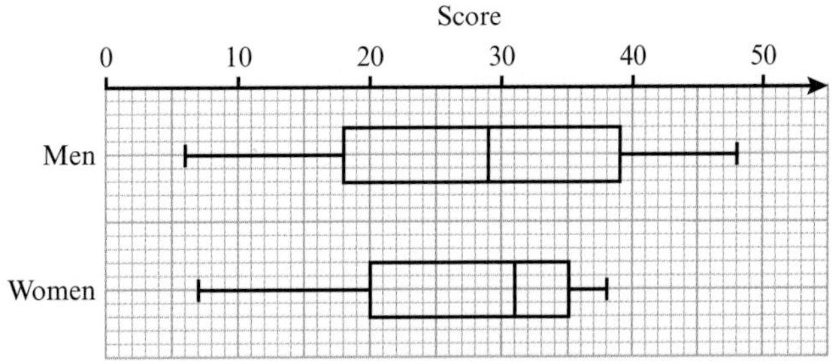

a For the men taking the aptitude test, state the value of:

i the median score

ii the range of the scores

iii the interquartile range of the scores.

b Compare briefly the scores obtained by men and women, stating one similarity and one difference.

3.3 Variance and standard deviation

WORKED EXAMPLE 3.2

Eunice has taken part in twenty 40 kilometre races in her career, but some of these she failed to complete. The grouped frequency table shows how many whole kilometres she completed in these races.

No. completed kilometres	26 – 30	31 – 35	36 – 40
No. races (f)	4	6	10

Calculate estimates of the mean and standard deviation of the distances she ran in these races.

Answer

The class given as 26–30 represents distances from 26 up to but not including 31 km, and so on.

The following table shows the data using class boundaries.

Distance run (d km)	**No. races (f)**	**Mid-values (x)**	xf	x^2f
$26 \leqslant d < 31$	4	28.5	114	3249
$31 \leqslant d < 36$	6	33.5	201	6733.5
$36 \leqslant d \leqslant 40$	10	38	380	14440
	$\Sigma f = 20$		$\Sigma xf = 695$	$\Sigma x^2 f = 24\,422.5$

Note that the upper boundary of the data is 40 because that is the greatest possible distance she can run in any race.

Mean, $\bar{d} = \dfrac{695}{20} = 34.75\,\text{km}$

Standard deviation $= \sqrt{\dfrac{24\,422.5}{20} - 34.75^2} = 3.68\,\text{km}$

TIP

The word 'completed' used in the question warns us to think carefully about class boundaries. Using incorrect boundaries will lead to incorrect class mid-values.

EXERCISE 3B

P 1 State the situation in which the numerical value of the standard deviation of a set of data:

a is equal to the numerical value of the variance

b is greater than the numerical value of the variance.

2 Find the standard deviation of the following data sets, using the formula:

$\sqrt{\frac{1}{n}\sum(x-\bar{x})^2}$

a 2, 1, 5.3, −4.2, 6.7, 3.1

b 15.2, 12.3, 5.7, 4.3, 11.2, 2.5, 8.7

3 Correct to the nearest 0.1 kg, the masses of five people in a fitness class are 70.8, 68.7, 73.2, 79.7 and 82.6 kg.

a Find the mean mass. Explain briefly why your answer is an estimate.

b Calculate an estimate of the standard deviation of the masses.

4 An employee paints a design on plates in a factory. At the end of each day the plates are inspected and some are rejected. The table shows the number of plates rejected over a period of 30 days.

No. rejects	0	1	2	3	4	5	6
No. days	18	5	3	1	1	1	1

Show that the standard deviation of the daily number of rejects is approximately equal to one-quarter of the range.

5 Fifty children were asked how many female relatives they have. The table shows that the children said they have 3, 4, 6 or r female relatives.

No. female relatives	3	4	6	r
No. children (f)	14	22	8	6

a Find the value of r, given that the mean number of female relatives is exactly 4.64.

b Use your answer to part **a** to find the variance, correct to 2 decimal places.

PS 6 Six wind turbines are to be erected in a straight line across the fields of a farm. The wind turbine closest to the farmhouse, which will be in-line with the turbines, will be erected 200 metres away, and there will be exactly 250 metres between one turbine and the next.

Find the mean and standard deviation of the distances from the turbines to the farmhouse.

7 The circumferences, c metres, of 400 cylindrical oil drums are summarised in the following table.

Circumference (c metres)	$1.84 \leqslant c < 1.94$	$1.94 \leqslant c < 2.16$	$2.16 \leqslant c < 2.66$
No. oil drums (f)	200	130	70

a Write down a suitable value that can be used to represent the circumference of the oil drums in each class.

b Use your values from part **a** to calculate an estimate of:

i the mean circumference

ii the variance of the circumferences.

8 The mass of coffee in each of 80 packets of a certain brand was measured, correct to the nearest gram. The results are shown in the table.

Mass (grams)	244–246	247–249	250–252	253–255	256–258
No. packets	10	20	24	18	8

a Estimate the mean and standard deviation of the masses, setting out your work in a table.

b State two ways in which the accuracy of these estimates could be improved.

9 The ages, in completed years, of the 104 workers in a company are summarised as follows.

Age (years)	16–20	21–25	26–30	31–35	36–40	41–50	51–60	61–70
Frequency	5	12	18	14	25	16	8	6

a Estimate the mean and standard deviation of the workers' ages.

b In another company, with a similar number of workers, the mean age is 28.4 years and the standard deviation is 9.9 years. Briefly compare the age distribution in the two companies.

10 The diagram shows a cumulative frequency graph for the lengths of telephone calls from a house during the first six months of last year.

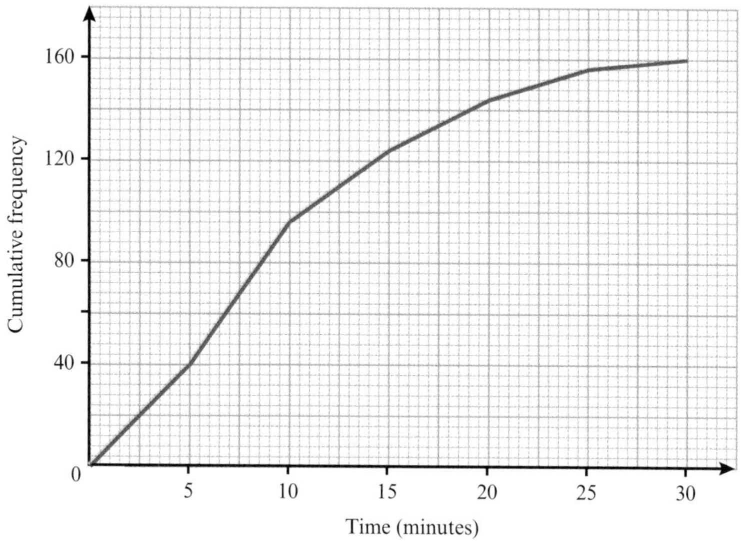

a Find the median and interquartile range.

b Construct a histogram with six equal intervals to illustrate the data.

c Use the frequency distribution associated with your histogram to estimate the mean length of telephone call.

d State whether each of the following is true or false.

i The distribution of these call times is skewed.

ii The majority of the calls last longer than 6 minutes.

iii The majority of the calls last between 5 and 10 minutes.

iv The majority of the calls are shorter than the mean length of call.

11 The heights of 94 male police officers based at a city police station were measured. The results (in metres) are summarised in the following table.

Height (m)	1.65–1.69	1.70–1.74	1.75–1.79	1.80–1.84	1.85–1.89	1.90–1.94	1.95–1.99	2.00–2.04	2.05–2.09
Frequency	2	4	11	23	38	9	4	2	1

a Draw a cumulative frequency diagram and estimate the median and quartiles.

b What do the values found in part **a** indicate about the shape of the distribution?

c Estimate the mean and standard deviation of the heights.

12 The following table gives the ages, in completed years, of the 141 people in a town involved in road accidents during a particular year.

Age (years)	12–15	16–20	21–25	26–30	31–40	41–50	51–70
Frequency	15	48	28	17	14	7	12

a Working in years, and giving your answers to 1 decimal place, calculate estimates of:

i the mean and standard deviation of the ages

ii the median age.

b Which do you consider to be the better representative average of the distribution, the mean or the median? Give a reason for your answer.

PS **13** A circular disc has three points, A, B and C, marked on its surface. These points are 4, 8 and 12 cm respectively from the centre of the disc.

a Find, in terms of π, the standard deviation of the distances through which the three points move during three complete revolutions of the disc.

b The standard deviation of the distances through which the points A, B and D move during $\sqrt{7}$ revolutions of the disc is 42π cm. Find the distance from the centre of the disc to point D.

Calculating from totals; combined sets of data

WORKED EXAMPLE 3.3

The times taken, to the nearest second, by the four fastest runners in a 1500 metre race are denoted by x. The times taken by the six slowest runners are denoted by y. These times are summarised by the totals:

$\Sigma x^2 = 249\,060$, $\Sigma x = 998$, $\Sigma y^2 = 428\,966$, $\Sigma y = 1604$

a Calculate the mean and standard deviation of the times for all of these ten runners together.

b Comment on the accuracy of the value that you have obtained for the mean.

Answer

a Mean $= \dfrac{\Sigma x + \Sigma y}{4+6} = \dfrac{2602}{10} = 260.2\text{ s}$

$$\text{SD} = \sqrt{\frac{\Sigma x^2 + \Sigma y^2}{4+6} - \left(\frac{\Sigma x + \Sigma y}{4+6}\right)^2} = \sqrt{\frac{678\,026}{10} - 260.2^2} = 9.93\text{ s}$$

b The mean of 260.2 is an estimate because the given totals are from times that have been rounded to the nearest whole number.

Lower boundary of mean $= \dfrac{996 + 1601}{10} = 259.7$

Upper boundary of mean $= \dfrac{1000 + 1607}{10} = 260.7$

$\therefore\ 259.7 \leqslant \text{mean} < 260.7$

$259.7 = 260.2 - 0.5$

$260.7 = 260.2 + 0.5$

The calculated value of 260.2 is correct to within 0.5 seconds.

Σx is the sum of 4 numbers, each rounded to the nearest integer, so 998 represents a class with boundaries $998 \pm (4 \times 0.5)$, i.e. $996 \leqslant \Sigma x < 1000$. Similarly, Σy is the sum of 6 numbers, each rounded to the nearest integer, so $1601 \leqslant \Sigma y < 1607$.

EXERCISE 3C

1 Find the standard deviation of x, given that 25 readings of x are summarised by the totals $\Sigma x = 490$ and $\Sigma x^2 = 15\,688$.

2 The masses, x grams, of the contents of 25 tins of Brand A anchovies are summarised by $\Sigma x = 1268.2$ and $\Sigma x^2 = 64\,585.16$. Find the mean and variance of the masses. What is the unit of measurement of the variance?

3 The standard deviation of ten values of x is 2.8. The sum of the squares of the ten values is 92.8. Find the mean of the ten values.

4 Find the value of Σy^2, given that 20 values of y have a mean of 12 and a variance of 80.

5 The following back-to-back stem-and-leaf diagram shows the masses of 20 female students and 18 male students.

	Female		Male	
(1)	8	4		(0)
(6)	8 8 5 4 4 2	5		(0)
(10)	9 7 6 6 5 4 3 3 2 1	6	0 1 5 5 7 9	(6)
(3)	9 2 0	7	0 8 8 8	(4)
(0)		8	1 1 2 2 4 5 7 7	(8)

Key: 1 | 6 | 5 means 61 kg for a female and 65 kg for a male

Summary: $\Sigma f = 1246 \quad \Sigma f^2 = 78\,704$

$\Sigma m = 1360 \quad \Sigma m^2 = 104\,162$

where f and m represent the masses of female and male students respectively.

Compare the masses of the females and males by drawing box-and-whisker diagrams and calculating the means and standard deviations of the masses.

6 25 values of p are such that $\Sigma p^2 = 5801$ and $\Sigma p = 380$.

25 values of q are such that $\Sigma q^2 = 6004$ and $\Sigma q = 385$.

a Confirm that the mean of the 50 values of p and q together is equal to $\frac{1}{2}(\bar{p} + \bar{q})$.

b Determine whether the variance of the 50 values of p and q together is equal to the mean of the variances of p and q.

7 **a** It is given that $\Sigma fx^2 = 5416$, $\Sigma f = 36$ and that the standard deviation of x is 9. Find Σfx.

b It is given that $\Sigma fx^2 = 198\,780$, $\Sigma fx = 3420$ and that the variance of x is 64. Find Σf.

8 The ages, correct to the nearest ten years, of 10 churches and 20 community halls in a city are denoted by x and y respectively. The ages are summarised by:

$\Sigma x^2 = 3\,918\,760$, $\Sigma x = 6270$, $\Sigma y^2 = 147\,920$, $\Sigma y = 1770$

a Find the mean and standard deviation of the ages of all these 30 buildings together.

b Comment on the accuracy of the value that you have obtained for the mean.

9 40 values of x are summarised by the totals $\Sigma x^2 = 4024$ and $\Sigma x = 380$.

n values of y are summarised by the totals $\Sigma y^2 = 8664$ and $\Sigma y = 522$.

The variance of all these values of x and y considered together is equal to approximately 19.277. Find the value of n.

10 The percentage score of an individual girl in an examination is denoted by x_g and the percentage score of an individual boy is denoted by x_b. For the 77 girls who took a particular examination, $\Sigma x_g = 5460$ and $\Sigma x_g^2 = 406\,530$. For the boys who took the same examination, $\Sigma x_b = 4965$. Given that the standard deviation for the scores of all 150 candidates was 14.5, find the value of Σx_b^2.

P **11** Five mothers have mean height 162 cm with standard deviation 7.2 cm. Each mother has a one-month old baby The mean height of the babies is 56 cm with standard deviation 3.8 cm.

Let the mean and standard deviation of the heights of this group of ten people together be denoted by m and s, respectively. Use inequalities to describe the possible value of m, and show that s > 50 cm.

P **12** A straight line passes through (0, –3) and (6, 0). The coordinates of ten randomly selected points on this line are denoted by (x, y).

a In the case where $\bar{x} = 2.8$ and $\Sigma y^2 = 42.5$, evaluate:

i $\bar{y}$ **ii** Σx^2.

b Use similar notation to describe a situation in which $(\bar{x}, \bar{y})$ is at $(2, -2)$.

Coded data

WORKED EXAMPLE 3.4

For eight values denoted by x, $\Sigma(x - 10)^2 = 20$ and $\Sigma(x - 10) = 12$, and for ten values denoted by y, $\Sigma(y - 3)^2 = 28$ and $\Sigma(y - 3) = 15$, find:

a the value of Σx^2

b the value of Σy^2

c the variance of the 18 values of x and y combined.

Answer

a $\bar{x} = \frac{12}{8} + 10 = 11.5$ and $\Sigma x = 8 \times 11.5 = 92$

The mean and sum of n values can be found directly from $\Sigma(x - b)$.

$$\text{Var}(x-10) = \frac{20}{8} - \left(\frac{12}{8}\right)^2 = 0.25$$

$$\text{Var}(x) = \frac{\Sigma x^2}{8} - 11.5^2 = 0.25$$

$\text{Var}(x) = \text{Var}(x-b)$

So $\Sigma x^2 = 8 \times (0.25 + 11.5^2) = 1060$

b $\Sigma(y - 3) = 15$

The mean and sum of n values can be found by expanding $\Sigma(y - b)$.

$$\Sigma y - \Sigma 3 = 15$$
$$\Sigma y - 30 = 15$$
$$\Sigma y = 45$$
$$\bar{y} = \frac{45}{10} = 4.5$$

'$\Sigma 3$' means 3 for each of the ten values of y, i.e. 30.

$$\text{Var}(y - 3) = \frac{28}{10} - \left(\frac{15}{10}\right)^2 = 0.55$$

$$\text{Var}(y) = \frac{\Sigma y^2}{10} - 4.5^2 = 0.55$$

$\text{Var}(y) = \text{Var}(y - b)$

So $\Sigma y^2 = 10 \times (0.55 + 4.5^2) = 208$

c Variance of x and y

$$= \frac{1060 + 208}{18} - \left(\frac{92 + 45}{18}\right)^2$$
$$= 12.5 \text{ (3 s.f.)}$$

Variance of x and y $= \frac{\Sigma x^2 + \Sigma y^2}{8 + 10} - \left(\frac{\Sigma x + \Sigma y}{8 + 10}\right)^2$

EXERCISE 3D

1 A set of ten data values, denoted by x, is such that $\Sigma x^2 = 850$ and $\Sigma(x - 1) = 80$. Evaluate $\bar{x}$ and find the variance of x.

2 The ages, in months, of 16 babies at a maternity clinic are denoted by m. The ages are summarised by the totals $\Sigma(m - 2)^2 = 312$ and $\Sigma(m - 2) = 70$. Find the standard deviation of the ages of the babies.

3 Ali measured the masses, x grams, of 12 cucumbers. He found that:

$\Sigma(x - 350) = 210$ and $\Sigma(x - 350)^2 = 6486$

Find the mean and standard deviation of the masses of the cucumbers.

4 The values, x, in a particular set of ten data items are summarised by:

$\Sigma(x - a) = 50$ and $\Sigma(x - a)^2 = 360$

a Find the variance of x.

b Given that $\bar{x} = 12$, determine the value of a and find Σx^2.

5 A chicken farmer fed 25 newborn chicks with a new variety of corn. The following stem-and-leaf diagram shows the weight gains of the chicks after three weeks.

36	9
37	6
38	4 5 6
39	3 3 7 9 9
40	2 3 7 8
41	0 2 6 6
42	3 5 7
43	2 4
44	5
45	1

Key: 39 | 3 means 393 grams

a Find the median weight gain and interquartile range.

The data may be summarised by:

$\Sigma(x - 400) = 192$ and $\Sigma(x - 400)^2 = 11894$

where x grams is the weight gain of a chick.

b Calculate the mean and standard deviation of the weight gains of the 25 chicks, giving each answer to the nearest gram.

c Chicks fed on the standard variety of corn had weight gains after three weeks with mean 392 grams and standard deviation 12 grams. State briefly how the new variety of corn compares to the standard variety.

6 In an experiment, 40 readings of x are summarised by the totals $\Sigma x^2 = 19\,600$ and $\Sigma x = 840$. Find the value of $\Sigma(x - 3)^2$.

7 The number of young in a brood of ducks is denoted by x. The following list gives values of $x - \bar{x}$ for 30 of these ducks, where $\bar{x}$ is the mean number of young in a brood.

2	1	0	4	−3	2	−3	0	1	−4
−3	4	1	3	−2	−1	−4	−4	0	0
3	−5	2	−4	−5	3	3	2	3	4

a Evaluate the totals $\Sigma(x - \bar{x})$ and $\Sigma(x - \bar{x})^2$.

b Given that $\Sigma x^2 = 3258$, find the mean and the variance of the number of young ducks.

8 The numbers of goals scored in 40 football matches are denoted by y. Values of y are summarised by the totals $\Sigma(y-3)^2 = 60$ and $\Sigma(y+3) = 195$. Find the value of Σy^2.

9 Each of 25 boys owns at least three pairs of trousers. The number of pairs of trousers that a boy owns is denoted by t. The table shows the frequencies for values of t–3.

$t-3$	0	1	2	3	4
No. boys (f)	2	5	9	6	3

a How many of the boys own four or fewer pairs of trousers?

b Find the mean and the standard deviation of the numbers of pairs of trousers owned by these boys.

M 10 Two men are going on a six-day walking trip and they plan to cover a total distance of 126 km.

a Find the mean distance they plan to walk each day.

b The variance of the daily distances they plan to walk is 6.25 km^2. However, due to poor map-reading skills, they actually walk 2 km further each day than planned.

i What is the mean distance that they actually walk each day?

ii Find the standard deviation of the actual daily distances they walk, and explain the reason behind your answer.

11 75 values denoted by x and 25 values denoted by y are summarised by the following totals:

$\Sigma(x+4)^2 = 4444$, $\Sigma(x+4) = 444$, $\Sigma(y+2)^2 = 2222$, $\Sigma(y+2) = 222$

a Calculate the value of $\bar{x}$ and of $\bar{y}$.

b Show that $\Sigma x^2 = 2092$, and find the value of Σy^2.

c Find the exact variance of the 100 values of x and y together.

12 120 children entered a talent competition. The number of marks awarded to each of the 65 girls is denoted by g, and the number of marks awarded to each of the boys is denoted by b. The marks awarded are summarised by the totals:

$\Sigma(g-50)^2 = 600$, $\Sigma(g-50) = 143$, $\Sigma(b-40)^2 = 5730$, $\Sigma(b-40) = 539$

a Show that the boys and girls together were awarded a total of 6132 marks.

b Calculate separately the sum of the squares of the boys' marks and the sum of the squares of the girls' marks.

c Find the standard deviation of the marks awarded to the girls and boys together.

E

WORKED EXAMPLE 3.5

For 15 values denoted by x, $\Sigma\left(\frac{1}{2}x - 3\right)^2 = 24$ and $\Sigma\left(\frac{1}{2}x - 3\right) = 9$. Find the value of Σx^2.

Answer

$$\bar{x} = \left(\frac{9}{15} + 3\right) \times 2 = 7.2$$

Use $\text{Var}(x) = \frac{1}{a^2} \times \text{Var}(ax - b)$, where $a = \frac{1}{2}$.

$$\text{Var}(x) = 4 \times \text{Var}\left(\frac{1}{2}x - 3\right)$$

$$\frac{\Sigma x^2}{15} - 7.2^2 = 4 \times \left[\frac{24}{15} - \left(\frac{9}{15}\right)^2\right]$$

$$\frac{\Sigma x^2}{15} - 7.2^2 = 4.96$$

$$\Sigma x^2 = 15 \times (4.96 + 7.2^2) = 852$$

Alternative method

In this second method we will use expansion of brackets.

$$\Sigma\left(\frac{1}{2}x - 3\right) = 9$$

$$\frac{1}{2}\Sigma x - \Sigma 3 = 9$$

$\Sigma 3$ means $15 \times 3 = 45$.

$$\frac{1}{2}\Sigma x = 9 + 45$$

$$\Sigma x = 108$$

$$\Sigma\left(\frac{1}{2}x - 3\right)^2 = 24$$

$$\Sigma\left(\frac{1}{4}x^2 - 3x + 9\right) = 24$$

$$\frac{1}{4}\Sigma x^2 - 3\Sigma x + \Sigma 9 = 24$$

$\Sigma 9$ means $15 \times 9 = 135$.

$$\frac{1}{4}\Sigma x^2 - (3 \times 108) + 135 = 24$$

$$\Sigma x^2 = 4 \times (24 + 324 - 135)$$

$$\Sigma x^2 = 852$$

EXERCISE 3E

1 On a hot day, a market trader puts some watermelon segments on his stall. The variance of their masses is 160 000 grams^2. After an hour, each segment has lost 0.05% of its mass due to water evaporation. Find the standard deviation of the masses after an hour. What assumption must you make for this calculation to be valid?

M **2** Nadia is going to invest some of her savings for five years in an account that pays compound interest at a rate of 7.5% per annum. The amount that she is thinking of investing will generate interest of $\$x$. By what percentage should she increase the amount that she invests so that the interest generated is $\$3x$?

3 A taxi can be hired by paying a fixed charge of \$4 plus \$1.65 per kilometre travelled. Find the difference between the amount paid by Myfanwy and Blodwyn, if Myfanwy travelled 7 km further in the taxi than Blodwyn.

4 At a funfair, Jessie buys three balloons which have been inflated such that their mean volume, in cm^3, is 20 000 with standard deviation 80.

a Write down the variance of the volumes. Include appropriate units with your answer.

b During the morning, the three balloons shrink at the same rate, so that the standard deviation of their volumes at lunchtime is 72 cm^3. Find the mean and variance of their volumes at lunchtime.

PS **5** A point on the line with equation $2y = 5 - x$ is translated by the vector $\begin{pmatrix} 6 \\ p \end{pmatrix}$ to a new position on the line. Find the value of p.

6 For 10 values denoted by x, $\Sigma(2x - 5)^2 = 90$ and $\Sigma(2x - 5) = 20$, find the value of Σx^2.

7 For 12 values denoted by x, $\Sigma\left(\frac{1}{3}x + 2\right)^2 = 2480$ and $\Sigma\left(\frac{1}{3}x + 2\right) = 120$, find the value of Σx^2.

8 For 20 values denoted by x, $\Sigma(5x - 7) = 4500$ and $\Sigma x^2 = 52\,000$, find the value of $\Sigma(5x - 7)^2$.

9 The average mass of a grain of a particular variety of rice after it has been boiled is 0.032 g with standard deviation 0.00192 g.

a Given that boiling the rice increases its mass by 28%, find the mean and variance of the mass of a grain before it is boiled.

b Joe boils 0.45 kg of rice and shares it equally between himself and the other three members of his family. Find the variance of the masses of rice on the diners' plates.

PS **10** Ten people invested various amounts of money, denoted by $\$x$, into Sofia's new business. After successfully running the business for a year, Sofia repays them $\$y$ each, where $25y - 33x = 12\,500$.

a Explain clearly how much each investor is repaid.

b Given that the amounts invested had a variance of 8 065 600 dollars^2, and that $\Sigma\left(\frac{1}{2}x - 10\,000\right) = 6000$, find the mean and standard deviation of the amounts received in repayment by the ten investors.

P

11 The mean and standard deviation of the radii of three circular discs are $\bar{r}$ and s metres, respectively.

a Find an expression in terms of $\bar{r}$ for $\bar{d}$, the mean diameter of the discs.

b Find an expression in terms of s for $\text{Var}(c)$, the variance of the discs' circumferences.

c Express the mean area of the discs in terms of $\bar{r}$ and s.

12 Temperatures can be measured using the Fahrenheit (F), Kelvin (K) or Celsius (C) scales, where $\text{K} = \text{C} + 273.15$ and $\text{C} = \frac{5}{9}(\text{F} - 32)$.

a Find, in the form $\text{F} = a\text{K} + b$, a formula for converting from Kelvin to Fahrenheit.

b Temperatures are measured at ten tropical locations and are summarised by the totals $\Sigma(\text{K} - 300)^2 = 500$ and $\Sigma(\text{K} - 300) = 70$. Find:

i the mean of these temperatures in Kelvin, in Celsius and in Fahrenheit

ii the values of ΣK^2, ΣC^2 and ΣF^2.

END-OF-CHAPTER REVIEW EXERCISE 3

1 The times taken, in minutes, for a large number of chemical processes to take place are represented in the cumulative frequency graph.

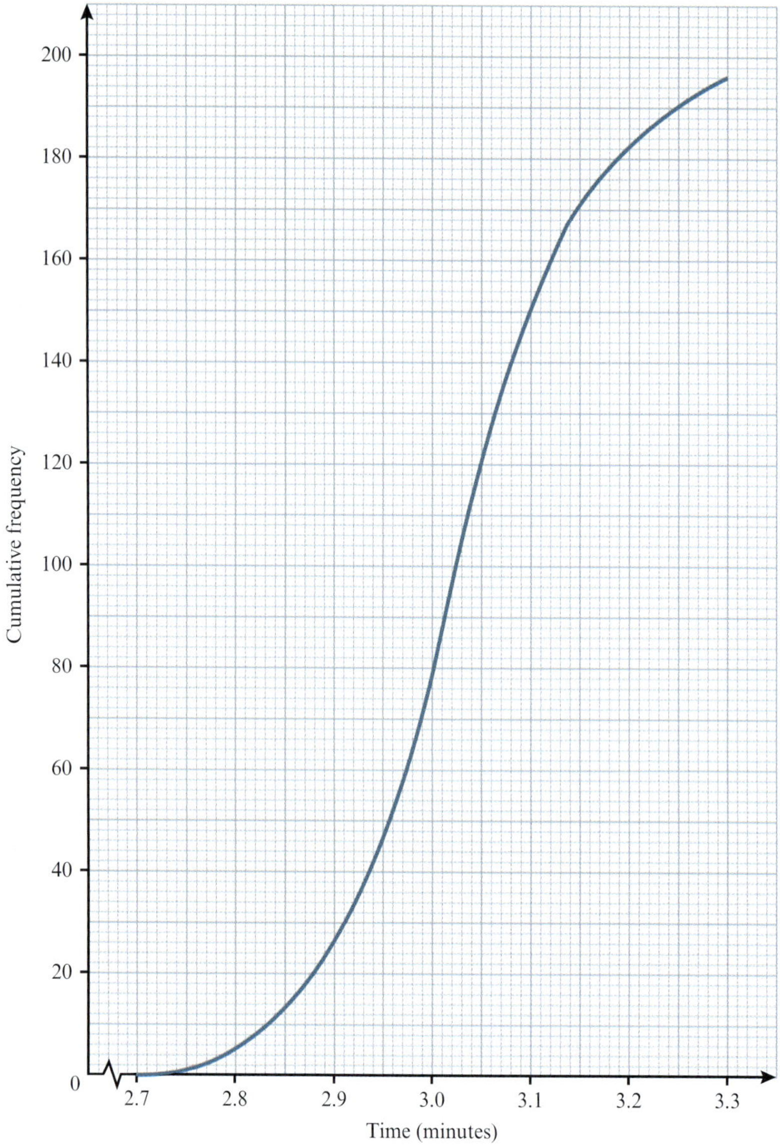

a How many chemical processes are represented in the graph?

b Estimate the median time taken.

c Estimate the number of processes that took:

i less than 3 minutes and 6 seconds

ii 177 seconds or more.

d Estimate the interquartile range of the times, giving your answer in seconds, correct to 1 decimal place.

2 Draw a box-and-whisker diagram to illustrate the data given in the cumulative frequency graph in question 1.

3 The wingspans, in centimetres, of seven mature robins and seven mature sparrows are given in the following table.

Robins	22.7	22.5	21.9	24.1	23.4	20.7	24.9
Sparrows	22.6	24.1	23.5	21.8	21.0	24.4	22.8

Find the mean and standard deviation of the wingspan of each type of bird, and use your results to compare the two sets of data.

4 The number of absences each day among employees in an office was recorded over a period of 96 days, with the following results.

No. absences	0	1	2	3	4	5
No. days	57	21	9	5	3	1

Calculate the mean and variance of the number of daily absences.

5 The times taken in a 20 km race were recorded for 80 runners. The results are summarised as follows.

Time (minutes)	60 –	80 –	100 –	120 –	180 – 200
No. runners	1	4	26	41	8

Calculate an estimate of the standard deviation of the times.

6 In a talent competition, the number of points awarded to a child competitor is denoted by x, and the number of points awarded to an adult competitor is denoted by y. A total of 32 children and 48 adults entered the competition and the scores awarded were such that $\Sigma x^2 = 142\,800$, $\Sigma y^2 = 200\,073$ and $\bar{y} = 63.0$. Find the mean number of points awarded to a child, given that the standard deviation for the number of points awarded to the children and adults together was 13.8.

7 For 40 values of x, it is given that $\Sigma x^2 = 232$ and $\Sigma x = 13.8$.

For 20 values of y, it is given that $\Sigma y^2 = 5\Sigma x$ and $\Sigma y = \frac{1}{4}\Sigma x^2$.

Calculate the standard deviation of the 60 values of x and y together.

8 46 readings of x have been coded and are summarised by $\Sigma(x - 3)^2 = 26.8$ and $\Sigma(x - 3) = 1.15$. Find:

a the value of Σx^2

b the smallest integer value of k for which $\Sigma(x + k) > 600$.

9 A group of four men and six women were asked how much, in dollars, they spent on magazines last week. The amount spent by an individual man is denoted by m and the amount spent by an individual woman is denoted by w. The amounts are summarised by the following totals:

$\Sigma(m - 5)^2 = 11.715$, $\Sigma(m - 5) = -3.8$, $\Sigma(w - 10)^2 = 82.66$, $\Sigma(w - 10) = 1.8$

a Find the mean amount spent on magazines last week by these ten people.

b By first finding the variance of m and the variance of w, show that $\Sigma m^2 = 73.715$ and find the value of Σw^2.

c Find, to the nearest cent, the standard deviation of the amounts spent by these ten people.

Chapter 4
Probability

- Evaluate probabilities by means of enumeration of equiprobable (equally likely) elementary events.
- Use addition and multiplication of probabilities appropriately.
- Understand the meaning of, and use the terms, mutually exclusive and independent events.
- Determine whether two events are independent.
- Calculate and use conditional probabilities.

4.1 Experiments, events and outcomes

Random selection and equiprobable events; exhaustive events; trials and expectation

WORKED EXAMPLE 4.1

A circular spinner is divided into five sectors with angles of 36°, 54°, 72°, 90° and 108°. The sectors are labelled, in the same order, with the numbers 2, 4, 6, 7, 8.

a Find the probability that an even number is obtained with each spin.

b How many times should we expect an odd number to be obtained with 60 spins?

Answer

a $\frac{360-90}{360}$ or $\frac{36+54+72+108}{360} = 0.75$

Even numbers occupy all but the 90° sector.

b $60 \times 0.25 = 15$

Event A is expected to occur $n \times \text{P}(A)$ times.

EXERCISE 4A

1 Find the probability that the number rolled with an ordinary fair die is:

a even

b not a 2

c a square number.

2 One letter is randomly selected from the word PARACHUTE. Find the probability that the letter is:

a an A

b in the word CANNOT.

3 A bag contains 20 coloured pencils: 2 are red, 5 are green, 6 are blue and the rest are grey. Find the probability that a randomly chosen pencil is:

a not green

b neither red nor blue.

4 An unfair coin is tossed 192 times and it lands on heads exactly 66 times. How many times should we expect the coin to land on tails when it is tossed 320 times?

5 Describe the single event that is complementary to each of the following.

a A 6 is not obtained when an ordinary fair die is rolled.

b Either Frank or Sabrina is selected to be the class representative.

c No goals are scored in the first half of tomorrow's Cup Final match.

6 One employee is randomly selected from a workforce that includes 40 females. The probability that a male is selected is 0.375. Find the probability that a particular female is selected.

PS 7 A rectangular grid contains 221 squares with rows headed 1, 2, 3, … and columns headed A, B, C, …

a Find the number of columns in the grid, given that it has more columns than rows and more than one row.

b Find the probability that a randomly selected square in this grid is:

i in the row headed 5

ii not in the column headed C.

PS 8 A clock, which has an hour hand and a minute hand only, is equally likely to stop at any time of day. The probability that the clock stops when the smallest angle between the hands is less than n degrees is equal to 0.3. Find the value of n.

4.2 Mutually exclusive events and the addition law

Venn diagrams

WORKED EXAMPLE 4.2

A fair 12-sided die with faces numbered from 1 to 12 is rolled. Find the probability that an even number or a factor of 20 is obtained.

Answer

Even numbers are {2, 4, 6, 8, 10, 12}, so $P(\text{even}) = \frac{6}{12}$

The events 'even' and 'factor of 20' are not mutually exclusive, because they have 3 common favourable outcomes.

Factors of 20 are {1, 2, 4, 5, 10}, so $P(\text{factor}) = \frac{5}{12}$

Even and a factor of 20 are {2, 4, 10}, so

$P(\text{even and factor of 20}) = \frac{3}{12}$

$P(\text{even or factor of 20}) = \frac{6}{12} + \frac{5}{12} - \frac{3}{12} = \frac{8}{12} = \frac{2}{3}$

	Factor of 20	Not factor of 20
Even	3	3
Not even	2	4

Alternatively, the solution can be found from a table showing the numbers of favourable outcomes.

$P(\text{even or factor of 20}) = \frac{3 + 3 + 2}{12} = \frac{2}{3}$

EXERCISE 4B

1 There is one boy named Ahmed in a class of 30 children that contains 12 other boys. A teacher selects one child at random. Find the probability that the selected child is:

a Ahmed

b not Ahmed

c a boy

d neither Ahmed nor a girl.

2 The Venn diagram shows some of the probabilities for events A and B.

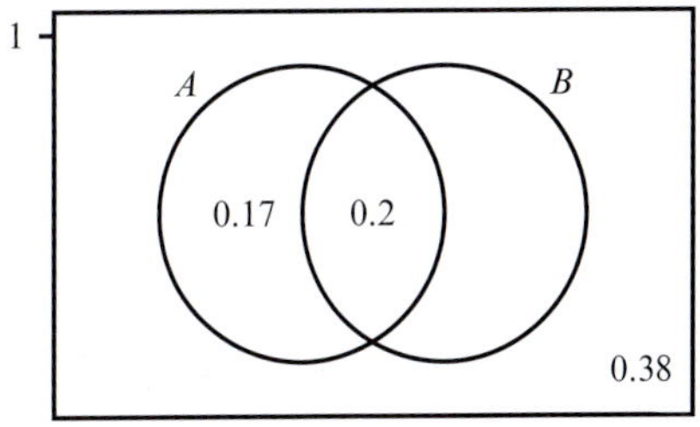

Find:

a $P(A')$

b $P(B)$

c $P(A \cup B)$.

3 An experiment has 14 equally likely outcomes. Seven outcomes are favourable to event X, eight outcomes are favourable to event Y and five outcomes are favourable to $X \cap Y$.

Use a Venn diagram, or otherwise, to find $P(X \cup Y)$.

4 The 32 students in a class are grouped by their gender and height in the following table.

	Short	Tall
Male	9	10
Female	6	7

a Find the probability that a randomly selected student is:

i not a tall male

ii short or female.

b Give a reason for what you notice about the answers to part **a i** and **a ii**.

5 Each of 50 performers was asked to state whether they had done any paid work in dance (D), mime (M) or acting (A) during the past year. Their responses are illustrated in the Venn diagram.

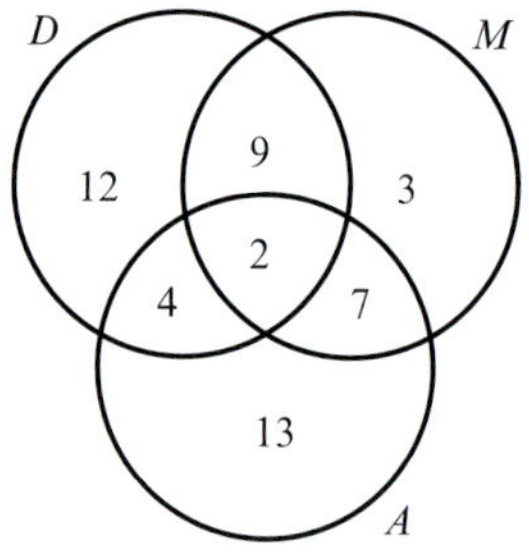

Find the probability that a randomly chosen performer had done paid work in:

a acting and dance

b acting and dance but not mime

c mime or dance

d dance or mime but not acting.

6 There are 20 patients in a dentist's waiting room. Eleven are there for a filling, eight for an extraction and three for neither of these reasons. By use of a Venn diagram, or otherwise, find the probability that a randomly selected patient is there for a filling or an extraction but not both.

7 Events M and N are such that $P(M) = 0.3$, $P(N) = 0.5$ and $P[(M \cup N)'] = 0.2$.

State, giving a reason, whether events M and N are mutually exclusive.

8 By use of a Venn diagram, or otherwise, find the probability that both or neither of the events X and Y occur, given that $P(X) = 0.45$, $P(Y) = 0.52$ and $P(X \cap Y) = 0.32$.

4.3 Independent events and the multiplication law

WORKED EXAMPLE 4.3

The events X and Y are independent, and it is given that $P(X) = 0.2$ and $P(X' \cap Y) = 0.24$. This information is shown on the tree diagram.

a Find the value of the probabilities labelled as a, b and c.

b Calculate $P(X \cap Y')$.

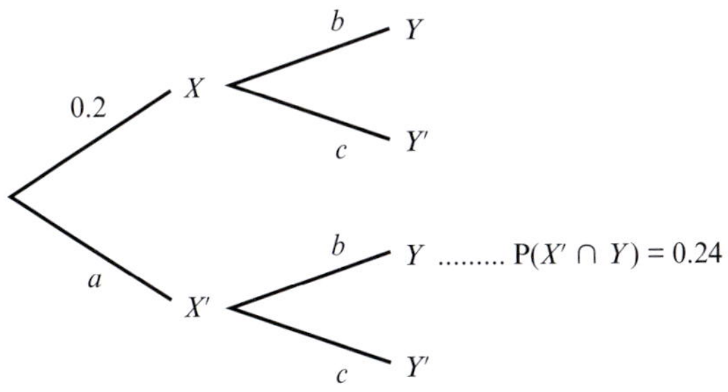

Answer

a $a = 1 - 0.2$
$= 0.8$

X' is the complement of X, so $P(X') = 1 - P(X)$.

$a \times b = 0.24$
$b = 0.24 \div 0.8$
$b = 0.3$

X and Y are independent, so X' and Y are also independent:
$P(X') \times P(Y) = P(X' \cap Y)$

$c = 1 - b$
$= 0.7$

Y' is the complement of Y, so $P(Y') = 1 - P(Y)$.

b $P(X \cap Y') = 0.2 \times c$
$= 0.2 \times 0.7$
$= 0.14$

X and Y are independent, so X and Y' are also independent:

$P(X \cap Y') = P(X) \times P(Y')$

EXERCISE 4C

1 A fair four-sided die has faces numbered 1, 2, 3 and 4. The die is rolled twice and the two numbers rolled are added together to give the total score.

a Copy and complete the following grid, which shows the total scores that can be obtained from the 16 equally likely events.

2nd roll					
	4			7	
	3				
	2			5	
	1	2			
		1	2	3	4
					1st roll

b Find the probability of obtaining a total score that is:

i equal to 5

ii less than 6

iii not more than 3.

2 A boy selects a number at random from 2, 3 and 4. A girl independently selects a number at random from 3, 4 and 5.

a The following tree diagram shows the probabilities for the two children selecting odd or even numbers. Copy and complete the diagram by writing in the missing probabilities.

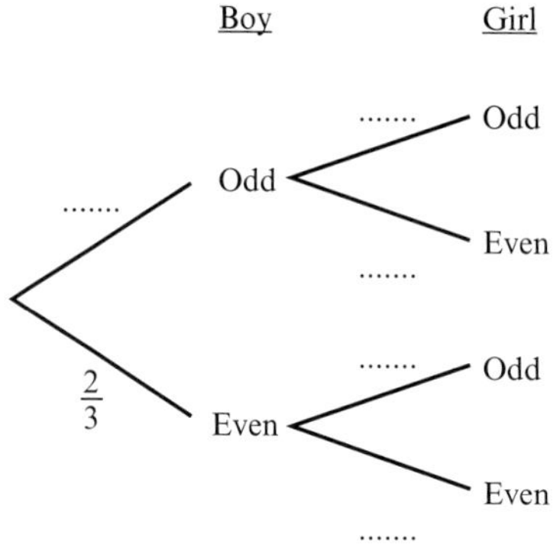

b Use your tree diagram to find the probability that:

i the boy and the girl both select odd numbers

ii the sum of the two numbers that they select is even.

3 A and B are two independent events. It is given that $P(A) = 0.7$ and $P(A \cap B) = 0.28$.

a Find $P(B)$.

b Copy and complete the following tree diagram by writing in the five missing probabilities.

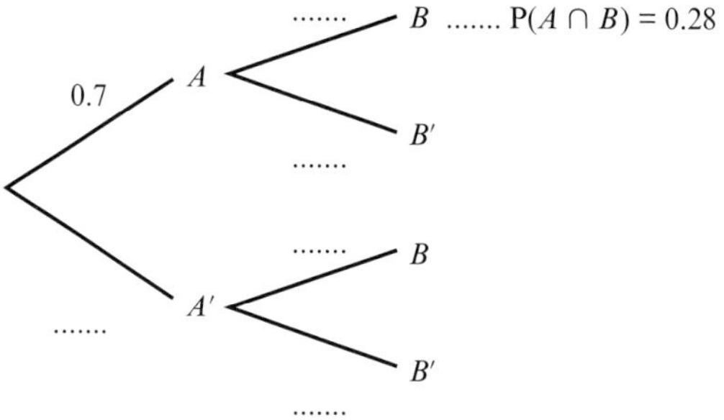

c Use your tree diagram to find $P(A' \cap B')$.

4 An experiment has 12 equally likely outcomes. Eight outcomes are favourable to event A, nine are favourable to event B, and $P(A \cup B) = 1$.

Use a Venn diagram, or otherwise, to find $P(A \cap B)$.

M 5 Calculate the probability that, in a randomly chosen family of four children, there are two males and two females. Assume that each child in a family is equally likely to be male or female, and that the gender of each child is independent of the gender of any previous children.

6 An ordinary fair die is thrown four times. Find the probability that:

a all four scores are 4 or more

b at least one score is less than 4

c at least one of the scores is a 6.

7 At a training college, 1 out of 36 students has red hair. The ratio of male to female students at the college is 5:6. Given that a student's hair colour is independent of their gender, find the probability that a randomly selected student is female and does not have red hair.

8 An ordinary fair die is rolled three times. Find the probability that:

a the product of the three numbers rolled is equal to 7 or 8

b the sum of the three numbers rolled is not equal to 3 nor 4 nor 5.

Application of the multiplication law

WORKED EXAMPLE 4.4

Each teenager in a group of 80 was asked whether or not they own a skateboard. Their responses are displayed in the following table.

	Boys (B)	Girls (B')	Totals
Owns skateboard (S)	12	8	20
Does not own skateboard (S')	36	24	60
Totals	48	32	80

One teenager is selected at random. Show that the event B (a boy is selected) and the event S (a skateboard owner is selected) are independent.

Answer

$\text{P}(B) = \frac{48}{80}$, $\text{P}(S) = \frac{20}{80}$

$\text{P}(B \cap S) = \frac{12}{80}$

We need to show that the multiplication law holds for events B and S.

$\text{P}(B) \times \text{P}(S) = \frac{48}{80} \times \frac{20}{80} = \frac{12}{80}$

So $\text{P}(B) \times \text{P}(S) = \text{P}(B \cap S)$

$\therefore$ Events B and S are independent.

TIP

Recall that $B \cap S$ has the same meaning as B and S.

EXERCISE 4D

1 Events V and W are independent. Find $P(V \cap W)$, given that $P(V) = 0.35$ and $P(W) = 0.24$.

2 Events X and Y are independent. Find $P(X \cap Y)$, given that $P(X') = 0.85$ and $P(Y') = 0.46$.

3 Events A and B are independent. Find $P(A')$, given that $P(B) = 0.48$ and $P(A \cap B) = 0.12$.

4 Independent events J and K are such that $P(J') = 0.7$ and $P(K') = 0.8$. Find:

a $P(J \cap K)$

b $P[(J \cup K)']$.

5 Two fair four-sided dice are each rolled once. The first die is numbered 2, 2, 3, 4 and the second die is numbered 1, 2, 3, 5.

Event A is 'the product of the two numbers obtained is greater than 6'.

Event B is 'the sum of the two numbers obtained is less than 7'.

a Show that $P(A \cap B) = \frac{1}{8}$.

b Explain how you know that the events A and B are not independent.

PS 6 The 69 items in a box are either black (B) or grey (G) and either wooden (W) or metallic (M). The number of each type is shown in the following table.

	W	M	Totals
B	14	11	25
G	28	16	44
Totals	42	27	69

a One item is randomly selected. Find the probability that it is grey and metallic.

b Show that the probability of selecting a grey item and the probability of selecting a metallic item are not independent.

c The probability of selecting a grey item and the probability of selecting a metallic item are independent if n black metallic items are removed from the box. Find the value of n, justifying your answer.

M 7 In a survey, 340 randomly selected people were each asked whether or not they wear spectacles and whether or not they can dance salsa. Their responses are shown in the table.

		Salsa		
		Can	Cannot	Totals
Spectacles	Yes	60	80	140
	No	80	120	200
Totals		140	200	340

a Determine whether the data support the theory that the events 'a person wears spectacles' and 'a person can dance salsa' are independent for this group of people.

b In the context of this question, give a generalised interpretation of your results from part **a**.

c Discuss the likelihood that the interpretation given in part **b** is true.

P 8 A student has 100 objects in a bag. The objects are categorised by their shape and colour. The table shows that 30 objects are square (S), 40 objects are black (B) and that x objects are both square and black.

	S	S'	Totals
B	x		40
B'			60
Totals	30	70	100

One object is selected at random from the bag.

a Given that the events 'a square object is selected' and 'a black object is selected' are independent, find the value of x.

b Show that the other three pairs of events (S and B', S' and B, S' and B') are also independent.

c Another bag contains T objects, of which s are square, b are black and x are both square and black. The information is shown in the following table.

	S	S'	Totals
B	x		b
B'			
Totals	s		T

Prove that if the events S and B are independent, then the other three pairs of events (S and B', S' and B, S' and B') are also independent.

4.4 Conditional probability

WORKED EXAMPLE 4.5

A group of 45 students from three different schools, X, Y and Z, were asked how many subjects they are studying at A Level. The results are given in the following table.

		School X	School Y	School Z	Totals
No. A Levels	2	6	5	7	18
	3	7	5	8	20
	4	4	1	2	7
Totals		17	11	17	45

One of the students is selected at random. Find the probability that this student:

a studies four A Levels, given that the student is from school Z

b is from school X, given that the student is studying fewer than four A Levels.

Answer

a $\frac{2}{17}$ — There are 17 students from school Z, and two of them study four A Levels.

b $\frac{6+7}{18+20} = \frac{13}{38}$ — There are 38 students who study fewer than four A Levels, and 13 of them are from school X.

EXERCISE 4E

1 The 11 digits of the number 'forty-five thousand, three-hundred and forty-five million, two-hundred and forty-five thousand, one hundred and forty-five' are written down. One of the digits is selected at random. Find the probability that:

a a 4 is selected, given that a 5 is not selected

b a 2 is selected, given that a digit less than 4 is selected.

2 Ben calls his mother on the telephone on one randomly selected day of each week. Find the probability that he calls her on a Monday or a Tuesday, given that he does not call her on a Wednesday.

3 Two different digits are selected from the nine digits 1 to 9. Find the probability that the second digit is odd, given that the first digit is:

a even

b odd.

4 Two fair coins are tossed. Find the probability that one head and one tail are obtained, given that two tails are not obtained.

5 A fair eight-sided spinner with sides numbered 2, 3, 3, 3, 5, 6, 7 and 7 is spun. Given that the number spun is a prime number, find the probability that it is an odd number.

6 The table shows the quiz marks of 150 people.

Marks	0	1	2	3	4	5	6	7	8	9	10
No. people (f)	1	4	5	11	13	26	32	29	14	7	8

One person is randomly selected from those who scored 3 or more marks. Find the probability that this person scored:

a fewer than 9 marks

b not fewer than 5 marks

c at least 1 more than the mean mark scored by the 150 people.

7 Three rectangular cards are arranged to create a 32 by 48 cm rectangle. The blue card measures 32 by 24 cm and the yellow card is a 24 cm square. The green card measures 12 by 48 cm and is partly covered by the blue card and the yellow card. Its left edge is indicated by a dotted line in the diagram.

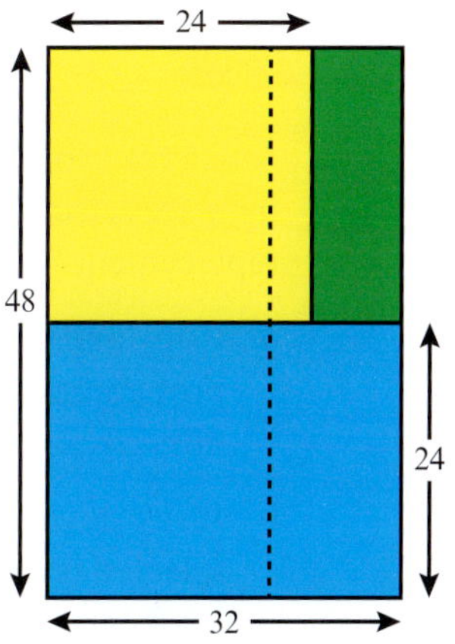

A dart is thrown and pierces a random point inside the 32 by 48 cm rectangle. Use areas to find the probability that the dart does not pierce the yellow card, given that it does pierce the green card.

8 Two fair dice, both numbered 1, 2, 3, 3, 4, 6, are rolled and the two numbers obtained are added together to give the score, X. Given that $X \leqslant 7$, find the probability that $X \geqslant 5$.

4.5 Dependent events and conditional probability

WORKED EXAMPLE 4.6

On the menu at an Italian restaurant there are 30 main course dishes. All of the dishes contain at least one of the herbs from basil (B), rosemary (R) and thyme (T). The numbers of dishes that contain these herbs are shown in the Venn diagram.

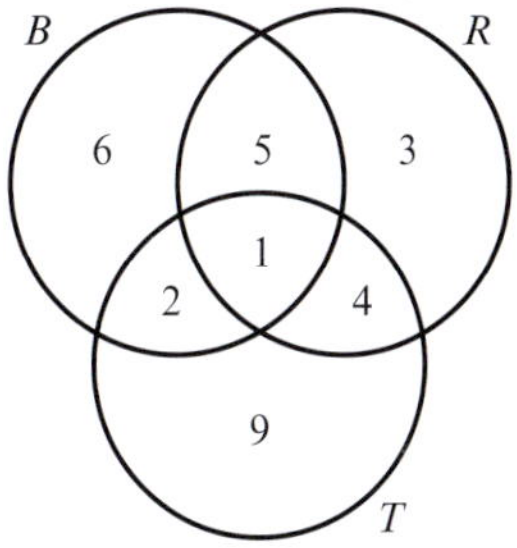

a Two different dishes are selected at random. Find the probability that they both contain thyme.

b Three different dishes are selected at random. Find the probability that exactly two of the dishes contain exactly one of the herbs.

Answer

a $\frac{16}{30} \times \frac{15}{29} = \frac{8}{29}$ or 0.276

The number of dishes that contain thyme is: $n(T) = 2 + 1 + 4 + 9 = 16$.

b $P(2 \text{ contain } 1 \text{ herb}) = P(H\,H\,H') + P(H\,H'\,H) + P(H'\,H\,H)$

$$= 3 \times \frac{18}{30} \times \frac{17}{29} \times \frac{12}{28}$$

$$= \frac{459}{1015} \text{ or } 0.452$$

$6 + 3 + 9 = 18$ dishes contain exactly one herb and 12 dishes do not contain exactly one herb. Let H represent the event 'a dish contains exactly one herb'.

EXERCISE 4F

1 A bag contains three apples and two pears. Two pieces of fruit are selected at random, one after the other and without replacement.

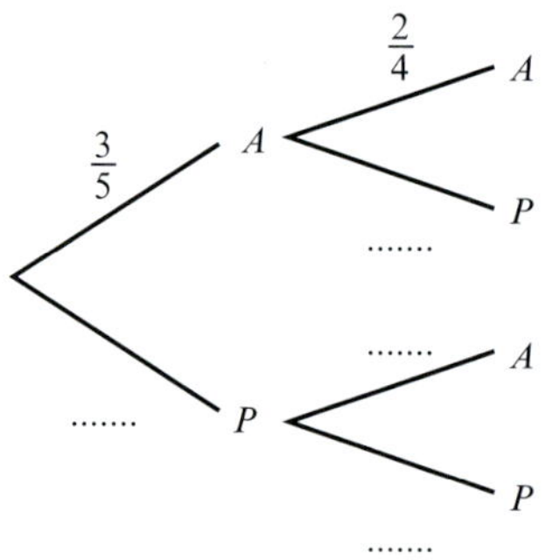

a Copy and complete the tree diagram, which shows the probabilities of the possible selections, by writing in the missing probabilities.

b Write down the probability that the second piece of fruit selected is an apple, given that the first piece of fruit selected is a pear.

c Find the probability that two apples or two pears are selected.

2 A drawer contains four identical black socks and three identical red socks. Two socks are selected at random from the drawer.

a Draw and fully label a tree diagram to show the probabilities for the possible selections.

b Use your diagram to find the probability that the selected socks:

i are both black

ii are both red

iii are not a matching pair.

3 a Given $P(A \cap B) = 0.252$ and $P(A) = 0.72$, find $P(B \mid A)$.

b Given $P(C \cap D) = 0.3034$ and $P(D \mid C) = 0.74$, find $P(C)$.

c Given $P(F \cap G) = 0.3478$ and $P(G') = 0.63$, find $P(F \mid G)$.

4 Two fair five-sided spinners, both with sides numbered 2, 3, 3, 5 and 7, are spun. The two numbers obtained are added together to give the score, X. Given that X is even, find the probability that the two numbers spun are the same.

5 A photographer takes portrait and landscape photographs in the ratio 1:2. The probability that she takes a landscape photograph that is not in focus is $\frac{22}{75}$, and the probability that she takes a photograph that is in focus is $\frac{8}{15}$. By use of a tree diagram, or otherwise, find the probability that a randomly selected photograph is a portrait, given that it is not in focus.

6 At the start of a game, a player's counter is placed on the zero of the following board.

0	1	2	3	4	5	6	7	8	9	10	11	12

A player rolls an ordinary fair die and moves her counter forwards a number of squares equal to the number rolled with the die. If the counter lands on a prime number (shown by the coloured squares), it is moved backwards two squares at a time, as many times as necessary for it not to land on a prime number. Then the die is rolled a second time and the counter is moved forwards as before.

Find the probability that after two rolls of the die a player's counter ends its move on:

a a prime number

b 0

c 8

d 1.

7 To qualify as a driver, a candidate must first pass a theory test and then pass a practical test. If a candidate passes the theory test, they are allowed two attempts to pass the practical test. If they fail the practical test at both attempts, they must start again by taking a new theory test.

The following table shows the proportion of candidates that pass these tests at their first, second, third, fourth, fifth and sixth attempts.

	1st	2nd	3rd	4th	5th	6th
Theory	65%	72%	80%	85%	91%	96%
Practical	35%	50%	65%	80%	95%	99%

a A particular candidate passes her first theory test. Find the exact probability that she has to take a theory test exactly once more before qualifying.

b Find the exact probability that a candidate qualifies on the fourth test that he takes.

8 A coin is biased so that tossing a head is twice as likely as tossing a tail. This coin is tossed and two identical four-sided dice, numbered 1, 2, 3 and 4, are rolled.

A player's score is calculated as follows.

- If the coin shows a head, the numbers rolled with the dice are multiplied together.
- If the coin shows a tail, the squares of the numbers rolled with the dice are added together.

Given that a player's score is 2, find the probability that the coin shows a tail if:

a the dice are fair

b the dice are biased, such that the probability of rolling a number is proportional to the number.

END-OF-CHAPTER REVIEW EXERCISE 4

1 An ordinary fair die is rolled. Find the probability that the number obtained is a multiple of 3 or a prime number.

2 Seven boys and six girls are standing in a line, with each girl standing between two boys. One boy is selected at random. Find the probability that he is standing between two girls.

3 Shinji has a collection of 40 books which he classifies as fantasy, mystery or crime. Some are paperbacks and some have hard covers, as shown in the table.

	Fantasy	Mystery	Crime
Paperbacks	10	12	10
Hard covers	2	1	5

a One of these books is selected at random. Find the probability that it is:

i a paperback mystery book

ii a fantasy book or a book with a hard cover

iii a crime book, given that it does not have a hard cover.

b Shinji now selects two fantasy books at random. Find the probability that at least one of them has a hard cover.

4 In his wardrobe, Moussa has 9 shirts: 4 are blue, 3 are white and 2 are red. He takes out two shirts, both selected at random.

a Find the probability that the second shirt he takes is red, given that the first shirt he took was:

i red **ii** not red.

b Find the probability that Moussa takes out two shirts of the same colour.

5 Students in a particular class were asked to represent a set of data in a diagram of their choice. From the 7 pie charts, 6 pictograms, 11 bar charts and 4 stem-and-leaf diagrams produced, the teacher selects at random four to pin onto a display board. Find the probability that the display has:

a four bar charts **b** no pie charts **c** exactly three pictograms.

PS **6** A computer randomly selects and arranges four digits from 0 to 9 in a square array. Any digit may appear any number of times. The top-left and bottom-right digits appear red, and the top-right and bottom-left digits appear blue. Two possible arrays are shown.

a Find the probability that, in any particular array:

i exactly one red zero appears

ii at least one zero appears.

$$\begin{bmatrix} 3 & 1 \\ 2 & 2 \end{bmatrix} \quad \begin{bmatrix} 6 & 7 \\ 5 & 0 \end{bmatrix}$$

b The product of the blue digits is subtracted from the product of the red digits to obtain a value denoted by D. For the examples in the diagram, $D = 4$ and $D = -35$. Find the probability that $D \geqslant 79$.

7 Forty students were given grades A, B or C for tests in Biology and in Chemistry. The numbers of students obtaining these grades are given in the following table.

		Chemistry		
		A	B	C
Biology	A	8	5	0
	B	3	10	2
	C	2	3	7

a Find the probability that a student with a grade A in Chemistry did not receive a grade A in Biology.

b A student with a grade B in Biology is selected at random, and then a different student with a grade B in Chemistry is selected at random. Find the probability that the second student selected received a grade B in Biology.

c For these particular students, 'obtaining grade P in X' and 'obtaining grade Q in Y' are mutually exclusive events. Indicate clearly the grades represented by the letters P and Q, and the subjects represented by the letters X and Y.

8 The probability that Dashiell goes to the gym and eats a burger on any evening is 0.2. The probability that he does not go to the gym and does not eat a burger is 0.48. The probability that he does not eat a burger on any particular evening is 0.68.

By use of a tree diagram, or otherwise, find the probability that Dashiell does not go to the gym, given that he eats a burger.

9 Two fair four-sided dice, one numbered 1, 2, 3, 4 and the other numbered 0, 1, 2, 3, are rolled simultaneously. The sum of the two numbers obtained is denoted by S.

A player who gets $S = 5$ then rolls a fair six-sided die and scores a number of points equal to the number rolled with this die.

A player who gets $S \neq 5$ scores 1 point.

Find the probability that a player scores:

a 1 point

b 1 point, given that the difference between the two numbers rolled with the four-sided die is 1.

10 The 225 men and 175 women in a group were each asked whether they had a bank card. 138 women said they did, and 41 men said they did not. One member of the group is selected at random. Showing your working, decide whether the events 'a woman is selected' and 'a person who has a bank card is selected' are independent.

Chapter 5
Permutations and combinations

- Solve simple problems involving selections.
- Solve problems about arrangements of objects in a line, including those involving repetition and restriction.
- Evaluate probabilities by calculation using permutations or combinations.

5.1 The factorial function

WORKED EXAMPLE 5.1

Without using a calculator, evaluate $\frac{6! - 3 \times 5!}{7!}$.

Answer

$\frac{6! - 3 \times 5!}{7!} = \frac{5!(6-3)}{7 \times 6 \times 5!} = \frac{3}{42} = \frac{1}{14}$ Factor out 5! from the numerator and denominator.

EXERCISE 5A

Do not use a calculator for this exercise.

1 Evaluate $\frac{1}{3!} \times 6!$

2 Evaluate $\frac{1}{2! \times 3!} \times 4!$

3 Evaluate:

a $9 \times 5! - 27 \times 4!$

b $\frac{15 \times 9! - 10 \times 8!}{5 \times 10!}$

4 Express, using factorials only, the total cost of seven cakes at \$8 each and six packets of biscuits at \$5 each.

5 A rectangle has an area of $(5! - 3!)\,\text{cm}^2$ and a width of $3!\,\text{cm}$. Find its length.

6 Express the number 4608 using factorials and indices only.

7 Show that $\frac{4!}{8!} = \frac{3!}{2! \times 7!}$.

PS 8 Solve the equation $\sin^2\theta = \frac{3!}{(2!)^3}$ for $0° \leqslant \theta \leqslant 360°$.

PS 9 The longest and shortest sides of a right-angled triangle have lengths of $\sqrt{7!}$ and $\sqrt{6!}\,\text{cm}$. Find the values of a and b, given that the length of its third side can be expressed as $a!\sqrt{b!}\,\text{cm}$.

5.2 Permutations

Permutations of n distinct objects

WORKED EXAMPLE 5.2

In how many different ways can the following be arranged in a line?

- **a** three elephants
- **b** five mice
- **c** three elephants and five mice

Answer

a $3! = 6$ ways — The 3 elephants are distinct objects.

b $5! = 120$ ways — The 5 mice are distinct objects.

c $8! = 40\,320$ — The 8 animals are distinct objects.

EXERCISE 5B

1 How many 7-digit numbers can be made from all of the digits 1, 3, 4, 6, 7, 8 and 9?

2 To celebrate the homecoming of their son, a family organises a six-course meal. In how many different orders can the courses be served?

3 In how many different ways can the following be placed in a line?

a six plant pots

b four vases of flowers

c six plant pots and four vases of flowers

4 From a pack of 52 cards, find how many ways there are of arranging in a row:

a all 26 red cards

b all 12 picture cards

c all but the 13 diamond cards.

5 One-hundred and twenty athletes start a marathon race. Exactly three-quarters of them drop out before the end. In how many different orders is it possible for those that complete the race to cross the finish line? You may assume that no two cross at exactly the same time.

PS 6 Find the smallest possible value of k such that half of the k books on a library shelf can be arranged in a row in more than 1 million million different ways.

PS **7** Seven boys and n girls can be arranged in a line in $3^4 \times 5^2 \times 7 \times 11 \times n^n$ different ways. Find the number of ways in which the girls alone can be arranged in a line.

M **8** Three modern sculptures are to be located on hilltops overlooking three towns, and each sculpture can be placed facing north, south, east or west.

At a meeting, the leaders of the town councils agree that this can be done in 24 different ways. Their thinking is that there are 3! ways of placing a sculpture on each hilltop, and four possible directions, so $3! \times 4 = 24$.

The artist who made the sculptures disagrees, saying that there are many more than 24 ways to arrange them. What error have the council leaders made in their calculations? Find the actual number of different ways that the sculptures can be placed.

Permutations of n objects with repetitions

WORKED EXAMPLE 5.3

How many 4-digit numbers can be made using the digits 5, 6, 7 and 8, if each digit can be used

a once

b any number of times?

Answer

a $4! = 24$ — Choices for 1st, 2nd, 3rd and 4th positions are 4, 3, 2 and 1.

b $4^4 = 256$ — Choices for 1st, 2nd, 3rd and 4th positions are 4, 4, 4 and 4.

EXERCISE 5C

1 Find the number of arrangements of all the letters in the following words.

a ABRACADABRA **b** ANTIDISESTABLISHMENTARIANISM

2 How many distinct 10-digit numbers can be made from the ten digits 1, 1, 2, 2, 3, 3, 4, 4, 5 and 5?

3 Four canaries and three love-birds are to be housed separately in seven cages. Explain why the number of ways in which this can be done is $7! = 5040$, and not $\frac{7!}{4! \times 3!} = 35$.

4 Daisy has written four poems and five short stories. All are to be published in a single book. In how many different orders is it possible to arrange these items in the book?

5 Four ordinary dice are placed in a row on a table.

a How many different arrangements of the four top numbers are possible?

b In how many of these arrangements are there four even numbers showing on top?

6 Two Xs and four other letters chosen from the remaining 25 letters of the alphabet are arranged in a row. Determine how many arrangements are possible in the following situations.

a The other four letters are all Zs.

b The other four letters are all the same.

c The other four letters are all different.

7 Six coloured beads (3 green, 2 red and 1 blue) are threaded onto a piece of string. Examples are shown in the diagram.

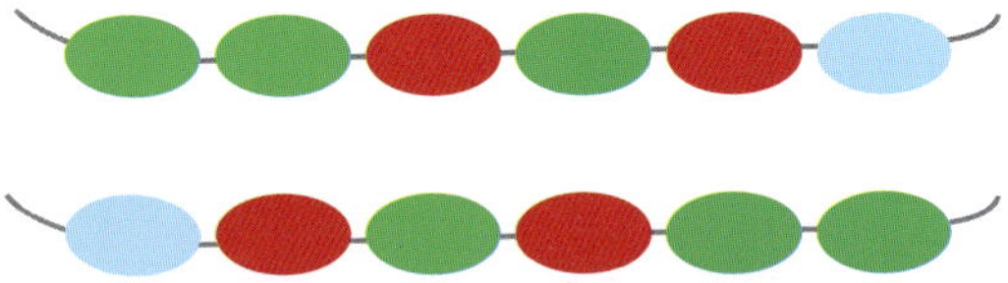

a Explain why the two examples shown are in fact identical arrangements of the three colours.

b In how many of the 30 distinct arrangements of the three colours are the two red beads separated?

M 8 Jason's 6-metre square kitchen floor, shown in the diagram, is to be covered using a pack of 12 tiles laid edge-to-edge. Each tile in the pack is a different colour and measures 1.5 by 2 metres.

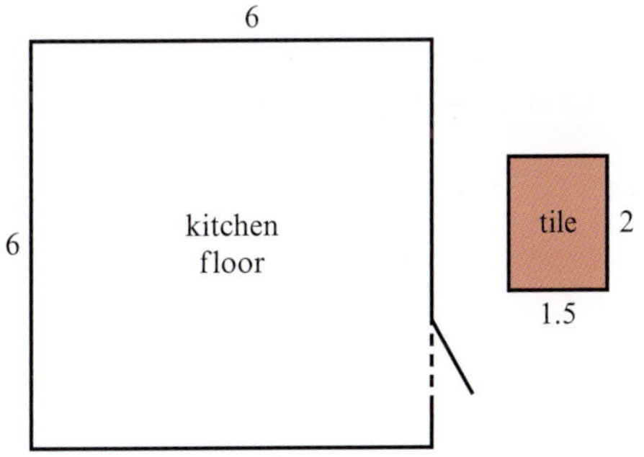

a How many different arrangements of the tiles are possible?

b The length and width of Jason's parents' kitchen floor are double those of his own.

i How many packs of these tiles must his parents buy to cover their kitchen floor?

ii Assuming that each pack of tiles is identical, and that all of the tiles are oriented the same way when laid, find the number of possible ways of tiling his parents' kitchen floor.

Permutations of *n* distinct objects with restrictions

WORKED EXAMPLE 5.4

A photograph is to be taken of seven girls standing in two rows, of three and four, with one row in front of the other. The tallest girl must be in the middle of the row of three, and the shortest girl must be in the row at the front. In how many ways can the girls be arranged?

Answer

The row of three can be at the front OR at the back, and we must look at these situations separately.

Let S represent the shortest girl and T represent the tallest girl (who is fixed in her position).

TIP

'OR' tells us to add the numbers of arrangements.

Three at the front

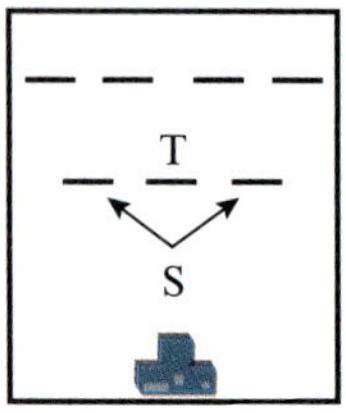

There are ${}^2P_1$ possible positions for S, AND there are ${}^5P_5$ ways to arrange the other five girls.

TIP

'AND' tells us to multiply the numbers of arrangements.

${}^2P_1 \times {}^5P_5 = 240$ arrangements

Three at the back

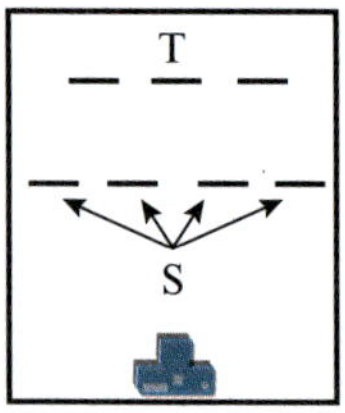

There are ${}^4P_1$ possible positions for S, AND there are ${}^5P_5$ ways to arrange the other 5 girls.

${}^4P_1 \times {}^5P_5 = 480$ arrangements

Totalling

$$\begin{aligned}\text{Total} &= ({}^2P_1 \times {}^5P_5) + ({}^4P_1 \times {}^5P_5)\\ &= 240 + 480\\ &= 720 \text{ arrangements}\end{aligned}$$

EXERCISE 5D

1 Find how many 4-digit numbers can be made using the digits 1, 2, 3 and 4 once each in the following cases.

  a There are no restrictions.

  b The 4-digit number must be:

    i less than 4000

    ii even

    iii less than 4000 and even.

2 In how many ways can nine toys be arranged in a line, if the largest toy must be in the middle of the arrangement?

3 A scout group of ten children and their two group leaders will walk in a line on parade, one behind the other. In how many ways can they do this, if the group leaders must be at the front and at the back of the line?

4 Find the number of different arrangements that can be made from all the letters in the word BANDANA, if:

  a there are no restrictions

  b the arrangement must begin with two Ns and end with three As.

5 Seven apples, five oranges and two pears are to be displayed in a row on a stall. Given that all the items are distinct, find how many ways they can be arranged, if:

  a there are no restrictions on their order

  b a pear must appear at both ends of the arrangement

  c the five oranges must not be separated.

6 Nine paintings are to be exhibited on the wall of an art gallery. They are to be hung in two rows with five in one row and four in the other. The brightest painting must be in the middle of the row of five, and the dullest painting must be at one end of the row of four. In how many ways can they be hung?

7 In the past year a boy has collected 12 editions of a monthly magazine, which he wishes to place in a neat pile on his bedside table. Determine how many ways he can do this, if:

  a January's edition is at the bottom of the pile and December's edition is at the top

  b neither January's edition is at the bottom of the pile nor is December's edition at the top

  c the editions for January and for December are next to each other.

8 Find the number of distinct arrangements that can be made from all the letters in the word BOMBARDED, if the arrangement:

a begins and ends with the letter B

b begins and ends with the same letter

c does not end with the letter M or the letter D.

PS 9 Eight grouped frequency tables are drawn up. Six histograms are drawn to represent the data from six of the eight tables. Students are asked, for 6 marks, to match each histogram with the grouped frequency table that it represents. Find the number of different ways for a student to:

a fail to score 6 marks

b fail to match just one histogram with the correct frequency table.

Permutations of r objects from n objects

WORKED EXAMPLE 5.5

Five letters are selected from the word ROTOVATOR and arranged in a row. How many of these arrangements begin with three Os?

Answer

There are 2 arrangements in which the last two letters are the same: RR or TT.

The arrangement is O O O _ _ and the last two letters can be the same or they can be different. Look at these two situations separately and add the numbers of arrangements together.

$2 + {}^4P_2 = 14$ arrangements

The last two letters are different when we select and arrange 2 from the 4 letters R, T, V and A.

EXERCISE 5E

1 Find how many permutations there are of:

a 3 from 13 distinct objects

b 17 from 19 distinct objects.

2 Find how many of the arrangements of five letters from A, B, C, D, E, F and G:

a end with the letter G

b do not begin with the letter B

c contain the letter F

d do not contain the letter C.

3 Mr Amu randomly selects 8 children from a group of 12. Mrs Batty then randomly dismisses half of these. The remaining children are taken to the school gym and arranged in a line. How many different arrangements of children could there be in the school gym?

PS 4 An insect walks along n of the 12 edges of a cube, each not more than once, from vertex A until it first arrives at vertex B.

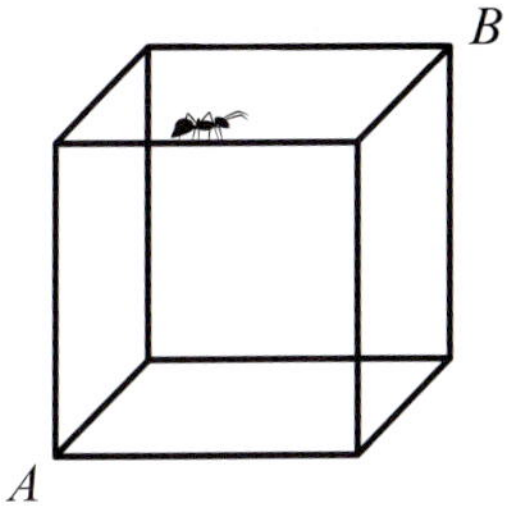

a How many different routes are possible when $n = 3$?

b List all the possible values of n.

5 Find the number of different ways that three letters from the word BANTAM can be arranged in a row, if the arrangement:

a must begin and end with the same letter

b cannot contain both of the As.

6 Using any of the digits 1, 2 and 3 any number of times, determine how many different n-digit numbers can be made when:

a $n = 2$ b $n = 1$ or 3.

PS 7 A group of six acrobats finishes its act in a triangular formation, as shown.

Find the number of different ways that the group can finish in this way, if the lightest acrobat must be at the top and the two strongest acrobats must have their feet on the ground.

PS 8 Over a continuous ten-day period, a salesman must make eight train journeys to visit clients in eight different towns. Because of the long distances, he can only visit one town on any particular day.

a In how many different ways can the salesman travel to these eight towns over the ten days?

b In how many ways can he arrange his travels, if he decides to visit clients on the first and the last of the ten days, and to have two consecutive travel-free days in between?

9 Five letters are selected from the word DECREASED and arranged in a row. How many of the possible arrangements end with two Ds?

5.3 Combinations

WORKED EXAMPLE 5.6

Gaston is planning to cook a stew. He decides to use exactly six ingredients from the five different types of beans and the seven different types of vegetables that he has in his kitchen.

a How many different combinations of six items does he have to choose from?

b In how many of these combinations are there more types of beans than types of vegetables?

Answer

a ${}^{12}C_6 = 924$

He chooses 6 items from 12.

b

Beans (5)	Vegetables (7)	
5	1	${}^5C_5 \times {}^7C_1 = 7$
4	2	${}^5C_4 \times {}^7C_2 = 105$
		Total = 112

He can use 5 types of beans and 1 type of vegetable, or he can use 4 types of beans and 2 types of vegetables.

The numbers of ways of doing this are tabulated.

There are more types of beans than types of vegetables in 112 of these combinations.

EXERCISE 5F

1 An examination paper consists of section A, which has six questions, and section B, which has eight questions. Find how many choices a candidate has, if they must answer:

- **a** two questions from section A or three questions from section B
- **b** two questions from section A and three questions from section B
- **c** any five questions.

2 Find the number of ways in which three girls can be selected from a group of:

- **a** seven girls
- **b** three girls and four boys.

3 **a** In how many ways can three silver coins be selected from ten silver coins?

- **b** In how many ways can four gold coins be selected from eight gold coins?
- **c** Find the number of ways in which three silver and four gold coins can be selected from a purse containing ten silver and eight gold coins.

4 From twenty letters and ten parcels, find how many ways there are to select:

a nine letters

b five parcels

c five letters and nine parcels.

5 **a** How many ways are there to select a committee of five people from a group consisting of five men and six women?

b Find how many of these committees consist of:

i two men and three women,

ii fewer men than women.

6 A team of seven is to be selected from the 15 members of a school chess club. Determine how many ways this can be done, if:

a the captain and the vice captain of the club must be selected

b either the captain or the vice captain, but not both, must be selected

c either the captain or the vice captain, but not both, can be selected.

7 A football team consists of 1 goalkeeper, 5 defenders and 5 attackers. Three members of the team are to be selected to attend a charity dinner. Find how many ways this can be done, if:

a the goalkeeper must be one of those selected

b at least one defender must be selected.

8 A team of 24 multi-talented athletes is to take part in four different events, where each of them may take part in one event only. Eight athletes are to be chosen for the sailing event, six for the decathlon, and five each for the long jump and high jump. Determine how many ways the athletes can be chosen, if:

a they are all willing to take part in any of the events

b 18 of them are not willing to take part in the decathlon

c the three tallest athletes must take part in the high jump.

P **9** Express n in terms of r, given that ${}^{n}C_{r+1} = {}^{n-1}C_{r}$.

5.4 Problem solving with permutations and combinations

WORKED EXAMPLE 5.7

A team of three people is selected at random from six girls and eight boys. Find the probability that the selected team consists of one girl and two boys.

Answer

No. possible combinations $= {}^{14}C_3$ — Select 3 from 14 people.

No. favourable combinations $= {}^{6}C_1 \times {}^{8}C_2$ — Select 1 from 6 girls and 2 from 8 boys.

$$\text{P(1 girl and 2 boys)} = \frac{{}^{6}C_1 \times {}^{8}C_2}{{}^{14}C_3}$$

$$= \frac{6}{13} \text{ or } 0.462$$

TIP

'Favourable' combinations contain one girl and two boys, as required by the question.

TIP

$$\text{Probability} = \frac{\text{No. favourable combinations}}{\text{No. possible combinations}}$$

WORKED EXAMPLE 5.8

A minibus has 12 seats: one for the driver (D) and 11 for passengers, as shown.

A group of four men and eight women has hired the minibus for a trip, but only two of the men and four of the women are qualified to drive it. Assuming that the driver and the seat occupied by each passenger are chosen at random, find the probability that the three seats at the front of the minibus are all occupied by women.

Answer

${}^{6}C_1$ or ${}^{6}P_1$ — Select a driver.

${}^{11}P_{11}$ — Arrange the 11 passengers.

No. possible permutations $= {}^{6}P_1 \times {}^{11}P_{11}$

$= 239\,500\,800$

${}^4C_1$ or ${}^4P_1$	Select a woman as driver.
${}^7P_2$	Arrange 2 of the remaining 7 women in the front seats.
${}^9P_9$	Arrange the remaining 9 passengers in the other seats.

$$\text{No. favourable permutations} = {}^4P_1 \times {}^7P_2 \times {}^9P_9 = 60\,963\,840$$

$$\text{P(3 front seats occupied by women)} = \frac{{}^4P_1 \times {}^7P_2 \times {}^9P_9}{{}^6P_1 \times {}^{11}P_{11}} = \frac{60\,963\,840}{239\,500\,800} = \frac{14}{55} \text{ or } 0.255$$

TIP

$$\text{Probability} = \frac{\text{No. favourable permutations}}{\text{No. possible permutations}}$$

EXERCISE 5G

1 A combination lock on a suitcase has three rotating discs, each numbered with the ten digits 0 to 9. Only one of the possible 3-digit numbers that can be made opens the suitcase. Find the probability that a randomly chosen 3-digit number fails to open it.

2 From a group of three men and two women, two people are to be selected at random to form a tennis doubles team.

- **a** In how many ways can the team be selected?
- **b** In how many ways can the doubles team consist of a man and a woman?
- **c** Find the probability that the two team members are not both women.

3 Four different letters are selected at random from the alphabet, which consists of 5 vowels and 21 consonants.

- **a** Find the probability that the selection consists of one vowel and three consonants.
- **b** Find the probability that, in a random arrangement of one vowel and three consonants, the vowel is not between two consonants.

4 At a meeting, everyone shakes hands once with everyone else.

a Find an expression in terms of n for the total number of handshakes that occur when there are n people at the meeting.

b At a particular meeting, there are 136 handshakes between two lawyers and 221 handshakes between a lawyer and a judge. How many handshakes are there between two judges?

5 Forty diners are to be seated in rows on five wooden benches. Two benches each have seats for five people, and three benches each have seats for ten people.

a In how many ways can two groups of five and three groups of ten be selected from the 40 diners?

b The two short benches are arranged so that the five diners selected for one bench are sitting back-to-back with the five diners selected for the other bench. Martina is to be seated on a particular short bench, but refuses to sit at either end; Gareth is to be seated on the other short bench, but refuses to sit back-to-back with Martina. In how many different ways can these ten diners be seated?

6 A simple calculator has 15 buttons. These are the ten digits from 0 to 9, the four basic operations, $+$, $-$, $\times$, $\div$, and an $=$ sign.

With the calculator screen showing 0, three buttons are pressed followed by the $=$ sign.

Find the number of valid calculations that can be performed if the first two buttons pressed are a digit and an operation (in either order) and the third button pressed is a digit.

7 A full pack of playing cards contains 52 cards, with 13 cards in each of the four suits: hearts, spades, diamonds and clubs. Anis selects four cards at random, and then Freida selects four cards at random from those that remain.

a Calculate the probability that both girls select one card of each suit.

b Find the probability that Anis selects more hearts than Freida.

8 Five noughts and four crosses are written into the nine squares of a grid. One of the possible arrangements is shown.

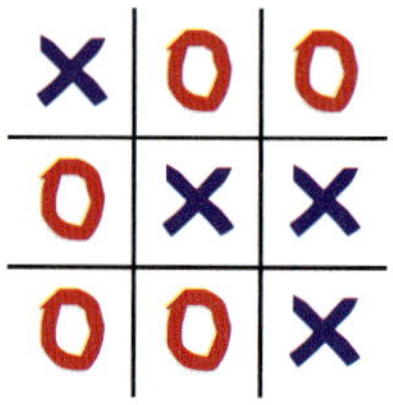

Given that all the noughts and crosses are written in randomly selected squares, find the probability that three crosses appear along any diagonal of the grid.

9 Claudio is planning an experiment, in which he will randomly select a direction from north, south, east and west, then walk five metres in that direction. He will do this n times altogether.

Event A is 'Claudio ends his journey where he started it'.

${}_n\mathrm{P}(A)$ denotes the value of $\mathrm{P}(A)$ for a particular value of n.

When $n = 2$, there are $4^2 = 16$ pairs of directions that Claudio can select, and four of these are favourable to event A, namely north south, south north, east west and west east.

So ${}_2\mathrm{P}(A) = \frac{4}{16} = \frac{1}{4}$

a Discuss how permutations can be used to find the number of favourable selections for $n = 4$, and show that ${}_4\text{P}(A) = \frac{9}{64}$.

b Tabulate the values of ${}_n\text{P}(A)$ for all values of n from 0 to 5 inclusive. Describe any patterns that you notice in the sequence of probabilities.

c Copy and complete the following table, which shows values of ${}_n\text{P}(A)$ for even values of n.

n	0	2	4	6	8	10
${}_n\text{P}(A)$		$\frac{1}{4}$	$\frac{9}{64}$			

d For even values of n, express ${}_{n+2}\text{P}(A)$ in terms of ${}_n\text{P}(A)$.

END-OF-CHAPTER REVIEW EXERCISE 5

1 A shop sells tubs of yoghurt in six flavours. Find how many selections of three tubs of yoghurt can be made, if:

a three different flavours must be selected

b exactly one strawberry flavoured yoghurt must be included.

2 Find how many arrangements of all the letters in the word ACCURATE can be made, if:

a there are no restrictions on the order

b the arrangement must not begin with the same two letters.

3 A construction company employs 20 people to work on three projects, which require groups of eight, seven and five workers. In how many ways can the 20 workers be assigned to the three groups?

4 The digits 1, 2, 3, 4 and 5 are to be used once each to create two numbers. One of the numbers must be greater than 100 and the other must be less than 100. Determine how many ways can this be done, if:

a there are no further restrictions

b the smaller of the two numbers is odd.

M 5 Three investment projects are presented to an investor who has \$30 000 to invest. Any investment that she makes must be in multiples of \$10 000. In how many different ways can the investor distribute her money if she uses all of it to invest in at least two of these projects?

6 A music teacher has offered free lessons to three girls and two boys. Each child may learn to play the cello, the violin or the clarinet.

a In how many different ways can the children choose the instrument they wish to learn?

b If each child chooses an instrument at random, find the probability that no two girls choose the same instrument, and that the boys choose the same instrument.

PS 7 During a football season, each team in a particular league plays each other team twice. Last season there were n teams in the league, but this season the number of teams is to be increased by 2. Given that the total number of games to be played this season is 70 more than last season, find the value of n.

PS 8 Axes for $-10 \leqslant x \leqslant 10$ and $-10 \leqslant y \leqslant 10$ are drawn and labelled. A point is marked at the origin (0, 0) and then a different point with whole-number coordinates is randomly selected and marked. A line segment is drawn between these two points.

a How many different line segments can be drawn?

b Find the probability that the line segment is not parallel to either of the axes.

9 Teenagers who stay overnight at a hostel are each given a task to perform in order to receive a free breakfast. There are 15 tasks, of which seven are to be done outdoors. The tasks are allocated at random to six girls and nine boys. Find the probability that more outdoor tasks are given to girls than to boys.

Chapter 6
Probability distributions

- Use a discrete random variable.
- Construct a probability distribution table relating to a given situation involving a discrete random variable X, and calculate its expectation, $E(X)$, and its variance, $Var(X)$.
- Use formulae for probabilities for the binomial and geometric distributions, and recognise practical situations where these distributions are suitable models.

6.1 Discrete random variables and 6.2 Probability distributions

WORKED EXAMPLE 6.1

The discrete random variable X is such that $X \in \{2, 4, 6, 8, 10\}$ and $P(X = x) = \frac{k}{x}$.

a Find the value of the constant k.

b Draw up the probability distribution table for X and find $P(X > 7)$.

Answer

a $\frac{k}{2} + \frac{k}{4} + \frac{k}{6} + \frac{k}{8} + \frac{k}{10} = 1$ — $\Sigma P(X = x) = 1$

$$\frac{120k + 60k + 40k + 30k + 24k}{240} = 1$$

$$274k = 240$$

$$k = \frac{120}{137}$$

b

x	2	4	6	8	10
$P(X = x)$	$\frac{120}{274}$	$\frac{120}{548}$	$\frac{120}{822}$	$\frac{120}{1096}$	$\frac{120}{1370}$

Find values of $P(X = x)$ by substituting for k and x in $\frac{k}{x}$.

$$P(X > 7) = P(X = 8) + P(X = 10)$$

$$= \frac{120}{1096} + \frac{120}{1370}$$

$$= \frac{27}{137}$$

EXERCISE 6A

1 The probability distribution for the discrete random variable X is given in the following table.

x	6	7	8	9	10
$P(X = x)$	$k - 0.1$	k	$1 - 3k$	$2 - 6k$	$\frac{1}{2}k + 0.05$

Find the value of the constant k.

2 The following table shows the probability distribution for Y.

y	1.0	1.1	1.2	1.3	1.4	1.5
$P(Y = y)$	a	0.14	0.27	0.18	0.08	b

a Find the value of $a + b$.

b Find $P(Y > 1.3)$, given that $a = 2b$.

3 The probability distribution for V is given in the following table.

v	3	4	5	6	7
$P(V = v)$	$k + 0.8072$	$0.2 - 2k$	$0.72k$	$0.1k$	k^2

a Show that the constant k satisfies the equation $k^2 - 0.18k + 0.0072 = 0$.

b Find the two solutions to the equation $k^2 - 0.18k + 0.0072 = 0$.

c Which of your two solutions to the equation $k^2 - 0.18k + 0.0072 = 0$ is valid in this question? Give a reason for your answer.

d Find $P(V \neq 4)$.

4 The discrete random variable T is such that $T \in \{2, 4, 6, 8\}$ and $P(T = t) = \frac{k}{t}$. Find the value of the constant k and state the value of T that is most likely to occur.

5 The discrete random variable Q is such that $Q \in \{1, 2, 3, 5\}$, and it is given that $P(Q = q) = \frac{1}{2} - kq$, where k is a constant.

a Show that $k = \frac{1}{11}$.

b Calculate the probability that Q is a factor of 6.

6 The probability distribution table for X is shown.

x	4	5	6	7	8
$P(X = x)$	0.3	0.34	0.18	0.12	0.06

a Use the information given in the probability distribution table to find the value of a, b and c that appear in the following cumulative probability distribution table for X.

x	3	4	5	6	7	8
$P(X \leqslant x)$	0	0.3	a	b	c	1

b Find $P(5 < X \leqslant 8)$.

7 On two days of every week, Hanna invites one of her three best friends to tea. On any day the probability that she invites Annie is 0.4, while Bettina and Chan are equally likely to be invited.

a Find the probability that Chan is not invited to tea on any particular day in a week.

b Draw up the probability distribution table for A, the number of times in any particular week that Annie is invited to tea.

8 A fair five-sided spinner has its edges labelled 1, 2, 3, 4 and 5. It is spun three times and the random variable X is the number of times that an odd number is obtained.

a Show that $P(X = 1) = \frac{36}{125}$.

b Draw up the probability distribution table for X.

c Find the value of $\frac{P(X = 2)}{P(X = 3)}$.

P **9** The random variable T is such that $T \in \{2, 3, 4, 5\}$. Show that there is no constant k for which $P(T = t) = 1 - kt$.

6.3 Expectation and variance of a discrete random variable

WORKED EXAMPLE 6.2

The probability distribution table for the random variable X is given. Find the mean and the variance of X.

x	4	5	6	7
$P(X = x)$	0.1	0.36	0.4	0.14

Answer

$E(X) = (4 \times 0.1) + (5 \times 0.36) + (6 \times 0.4) + (7 \times 0.14)$
$= 5.58$

The mean of a random variable is its expectation, $E(X) = \Sigma xp$.

$\text{Var}(X) = (4^2 \times 0.1) + (5^2 \times 0.36) + (6^2 \times 0.4) + (7^2 \times 0.14) - 5.58^2$
$= 0.7236$

$\text{Var}(X) = \Sigma x^2p - \{E(X)\}^2$

EXERCISE 6B

1 A keypad has ten buttons, each with a different single digit number from 0 to 9 on it. A girl is asked to randomly select and press a button. Find the expectation.

2 The edges of one fair triangular spinner are labelled with the numbers 1, 5 and a. The edges of another fair triangular spinner are labelled with the numbers 2, 6 and b.

a Find the value of a and of b, given that the expected scores with the spinners are 5 and 6 respectively.

b Each spinner is spun once and the two numbers obtained are added together to give the total score, T.

i Draw up the probability distribution table for T.

ii Use your table to find $E(T)$ and $Var(T)$.

iii State the probability that T is an even number.

3 A box contains four plain biscuits and two chocolate biscuits. Two biscuits are randomly selected, without replacement, from the box.

a Show that the probability of selecting two chocolate biscuits is $\frac{1}{15}$.

b i Draw up the probability distribution table for C, the number of chocolate biscuits selected.

ii Show that $E(C) = \frac{2}{3}$ and explain, without calculating any further probabilities, why the expected number of plain biscuits must be equal to $\frac{4}{3}$.

4 The probability distribution table for R is shown.

r	27	33	a	57
$P(R = r)$	0.2	0.3	0.4	0.1

Given that $E(R) = 38.2$, find the value of a and calculate $Var(R)$.

5 The probability distribution table for Y is shown.

y	m	$m + 1$	$m + 2$
$P(Y = y)$	0.2	0.5	0.3

Find, in terms of the constant m, an expression for $E(Y)$, and show that $Var(Y)$ is independent of the value of m.

M 6 A company needs to generate at least \$25 000 in its first year of trading to make a profit in a new line of business. A consultant has studied the market and produced the following table, which shows the estimated probability distribution of the amount generated in the first year.

Amount generated (\$)	Probability
–50 000 to –25 000	0.05
–25 000 to 0	0.20
0 to 35 000	0.30
35 000 to 65 000	0.30
65 000 to 85 000	0.10
85 000 to 100 000	0.05

Investigate the expectation and the variation in the amount generated. Advise the company about entering the new line of business. How confident are you about the advice that you offer?

7 In the first part of a game, a player tosses two fair coins. In the second part, the player rolls two fair dice, one fair die or none, depending on the result of the coin toss, as follows.

- A player who tosses two heads then rolls two fair dice, and is awarded a score equal to the absolute numerical difference between the numbers rolled.
- A player who tosses exactly one head then rolls one fair die, and is awarded a score equal to the number shown on the die if that number is odd, but equal to half that number if it is even.
- A player who tosses two tails is awarded a score of zero.

a Show that the probability of obtaining a score of 1 is equal to $\frac{17}{72}$.

b Find the probability of obtaining a score of not more than 1.

c Draw up the probability distribution table for the scores and find, correct to 2 decimal places, the standard deviation.

8 A case contains 12 identical memory sticks. However, seven of these are infected by a virus. Hassan takes two memory sticks at random from the case and passes it to Kara, who then takes three of the remaining memory sticks, also at random.

a Draw up the probability distribution tables for H and for K, the number of infected memory sticks taken by Hassan and by Kara.

b Who is expected to select the greatest number of infected memory sticks, Hassan or Kara? Justify your answer.

P 9 Two athletes, a sprinter and a hurdler, will each participate in two races this month. It is estimated that the sprinter has a 40% chance of winning any race, and that the hurdler independently has a 70% chance of winning any race.

a Draw up two probability distribution tables. One each for S and H, the number of races won by the sprinter and by the hurdler this month.

b The random variable W is the total number of races won this month by the sprinter and the hurdler together. Use your tables from part **a** to show that $P(W = 1) = 0.1944$.

c Draw up the probability distribution table for W.

d Verify that $E(W) = E(S) + E(H)$.

e Show how $E(W)$ can be calculated without using any probability distribution tables.

P **10** Six children are randomly selected from a group of ten that consists of two boys and eight girls.

a Draw up the probability distribution table for B, the number of boys selected.

b A number k of additional girls join the group of ten. Then six children are again selected at random. For all values of k, the ratio of the probabilities that no boys, one boy and two boys are selected can be expressed as $(k+4)(k+3):p(k+4):q$, where p and q are constants.

i Find the value of p and of q.

ii Show that when $k = 6$, the probability that one boy is selected is 0.5.

iii Find the value of k for which the probability that no boys are selected is 0.5.

PS **11** A box contains 14 tokens. Eight are worth \$5 each, four are worth \$3 each, and two are worth \$1 each. Rob selects four tokens at random from the box. The following table shows the ordered probability distribution for the total value of the tokens that he selects.

Value (\$)	a	$a+2$	$a+4$	$a+6$	$a+8$	$a+10$	$a+12$
Probability	$\frac{k-995}{k}$	$\frac{40}{k}$	$\frac{125}{k}$	$\frac{256}{k}$	$\frac{280}{k}$	$\frac{224}{k}$	$\frac{70}{k}$

a State the value of a.

b Calculate the value of k, and find the expected total value of Rob's four tokens, giving your answer correct to the nearest cent.

c Find the probability that the fourth token that Rob selects is worth \$5, given that his four tokens are worth a total of \$16.

M **12** The management of a hospital has produced the following table, which shows the probability distribution for the numbers of patients admitted each day with serious injuries.

No. serious injuries	< 5	6–10	11–15	16–20	> 20
Probability	0.08	0.4	0.46	0.05	0.01

Give some examples of how the information in the table could be used for the benefit of management, staff and patients.

PS **13** The random variable V is such that: $V \in \{14, 22, 34, 54, k\}$ and $P(V = v) = \frac{b - v^2}{11\,404}$

where b is a constant.

a Given that $b - 64k = 32$, find the value of b and of k.

b Draw up the probability distribution table for V, and show that the standard deviation of V is a little less than $\frac{1}{2}\overline{V}$.

END-OF-CHAPTER-REVIEW EXERCISE 6

1 The probability distribution table for the random variable X is given.

x	0	1	2	3
$P(X = x)$	$1-2k$	$2-4k$	$3-6k$	$4-8k$

a Find the value of the constant k.

b Show that $E(X) = 2$.

c Find $Var(X)$.

2 The table shows the probability distribution for W.

w	0	1	2	3
$P(W = w)$	$6p$	$2p$	$3p$	$4p$

a Find the value of the constant p.

b Find the mean of W and its standard deviation.

3 The discrete random variable Y is such that $Y \in \{a, a+2, a+4, a+6\}$. It is given that $P(Y = y) = ky$, where k is a constant.

a Express k in terms of a.

b Find $E(Y)$ when $a = 5$.

P 4 The probability of tossing a head with a particular coin is p. The coin is tossed on two occasions and the discrete random variable H is the number of heads obtained. The probability distribution table for H is given as follows.

h	0	1	2
$P(H = h)$	$(1-p)^2$	$2p(1-p)$	p^2

a Find an expression for $E(H)$ in terms of p.

b Show that $Var(H)$ can be expressed as $2p(1-p)$.

5 A bag contains seven balls that are numbered consecutively from 2 to 8. The random variable X is the number of even-numbered balls that are selected when two balls are taken at random from the bag.

a Draw up the probability distribution table for X.

b Find $E(X)$ and $Var(X)$.

c The random variable Y is the number of odd-numbered balls selected. Show that $E(Y)$ and $E(X)$ are different, but that $Var(Y)$ and $Var(X)$ are identical.

6 Benoit will take IGCSE exams in three science subjects this year. By assuming that subject grades are obtained independently, his three science teachers have created the following probability distribution table. It shows the likelihood of Benoit obtaining grade As.

No. grade As	0	1	2	3
Probability	0.09	a	b	0.14

a Form and solve a pair of simultaneous equations in a and b, given that the expectation for the number of grade As that Benoit will obtain in his science exams is 1.6.

b Find the probability that Benoit obtains a grade A in Chemistry, given that the probability that he obtains a grade A in Biology and in Physics is 0.35.

7 Two boxes each contain three chocolate bars and four fruit bars. Marie-Claire randomly selects, takes out and eats a bar from one of the boxes. She then puts all of the remaining bars into a bag which she offers to her friend. The friend takes two bars from the bag, both selected at random.

Let F be the number of fruit bars that the friend takes.

a Show that $P(F = 0) = \frac{15}{91}$.

b Draw up the probability distribution table for F.

c Hence, find the variance of F.

PS 8 The probability distribution for the random variable T is given in the table, where k is a constant.

t	$\frac{1}{4}k$	$\frac{1}{3}k$	$\frac{1}{2}k$	k
$P(T = t)$	k	$\frac{1}{2}k$	$\frac{1}{3}k$	$\frac{1}{4}k$

Find the standard deviation of T.

distributions

- Use formulae for probabilities for the binomial and geometric distributions, and recognise practical situations where these distributions are suitable models.
- Use formulae for the expectation and variance of the binomial distribution and for the expectation of the geometric distribution.

7.1 The binomial distribution

WORKED EXAMPLE 7.1

The random variable X follows a binomial distribution with $n = 9$ and $p = 0.25$. Find $P(X < 3)$, correct to 3 significant figures.

Answer

$X \sim B(9, 0.25)$ — Define the distribution.

$$P(X < 3) = P(X = 0) + P(X = 1) + P(X = 2)$$

$$= \binom{9}{0} \times 0.25^0 \times 0.75^9 + \binom{9}{1} \times 0.25^1 \times 0.75^8 + \binom{9}{2} \times 0.25^2 \times 0.75^7$$

$$= 0.075084\ldots + 0.22525\ldots + 0.30033\ldots$$

$$= 0.601$$

TIP

To achieve the accuracy required, work with 4 significant figures or more.

EXERCISE 7A

1 Given that $X \sim B(2, 0.84)$, find correct to 3 significant figures the value of:

a $P(X = 1)$

b $P(X \neq 1)$.

2 Given that $X \sim B\left(4, \frac{4}{7}\right)$, find the exact value of:

a $P(X = 2)$

b $P(X \leqslant 1)$.

3 The probability of tossing a head with a biased coin is 0.56. Find, correct to 3 significant figures, the probability that five tosses of this coin result in:

a exactly four heads

b exactly two tails

c more heads than tails.

4 A discrete random variable is: $X \sim \mathrm{B}\left(n, \frac{1}{2}\right)$. It is given that: $\mathrm{P}(X=0) = \mathrm{P}(X=n) = \frac{1}{2^n}$

Find expressions in terms of n for $\mathrm{P}(X=1)$ and for $\mathrm{P}(X=2)$, ensuring that neither of your answers contain factorials.

5 Two variables are $X \sim \mathrm{B}(4, 0.6)$ and $Y \sim \mathrm{B}(3, 0.4)$.

a Find the value of $\mathrm{P}(X=1)$ and of $\mathrm{P}(Y=1)$.

b State the condition under which $\mathrm{P}(X=1 \text{ and } Y=1) = \mathrm{P}(X=1) \times \mathrm{P}(Y=1)$.

c Assuming the condition stated in your answer to part **b** is met, find $\mathrm{P}(X+Y=2)$.

6 Tom tosses a fair coin five times and Aisha rolls five ordinary fair dice. Calculate the probability that together they obtain two or more heads and fewer than two sixes.

7 A new type of suitcase that can be pulled along on two wheels was tested. During the first hour of the test, 62% of suitcases lost their left wheel, 46% lost their right wheel and 12% lost neither of their wheels. Eight tested suitcases are then selected at random. Find the probability that fewer than three of them lost both wheels in the first hour of the test.

8 Independent trials are conducted in which the probability of success in each trial is 0.36. Find the least number of trials that must be conducted so that the probability of at least one success is greater than 99.5%.

P 9 A random variable is $X \sim \mathrm{B}(8, p)$. Find the value of $\mathrm{P}(X=5)$, correct to 4 significant figures, given that $\mathrm{P}(X=4) = 5 \times \mathrm{P}(X=2)$.

M 10 On a particular day, ten lambs are born on a sheep farm. Let F denote the number of female lambs born.

a State two assumptions you need to make so that F can be modelled by the distribution $\mathrm{B}(10, 0.5)$.

b Under these assumptions, find the probability that more than seven of the ten lambs born are females.

M 11 At a certain railway station, it is known that on average the probability that a train arrives late is 0.03. Twelve trains are due to arrive between 06:00 and 09:00 on a particular Monday morning. Let X denote the number of these trains that arrive late.

a In order to model X by a binomial distribution, it is necessary to assume that the probability of arriving late is the same for every train. Is this assumption reasonable? Give a reason for your answer.

b State a further condition required for X to be well modelled by a binomial distribution. Would it be reasonable to assume that this condition is met in this situation? Give a reason for your answer.

P **12** Over a period of two months, Zoe and Yifan play a total of n games of chess. The probability that Yifan does not lose each game is 0.2. The result of each game played is independent of any other game played. Let Y denote the number of games not lost by Yifan over this period.

a Show that $P(Y=2)$ can be given as $\frac{n(n-1)}{32} \times 0.8^n$.

b The probability that Yifan does not lose exactly two games is 0.268, correct to 3 significant figures. Evaluate n and find the probability that Yifan loses exactly ten of these games.

TIP

Use trial and improvement to answer the first part of **b**.

Expectation and variance of the binomial distribution

WORKED EXAMPLE 7.2

The random variable T follows a binomial distribution. Given that $E(T) = 25.52$ and $Var(T) = 3.0624$, find the probability that T takes a value that is greater than 27.

Answer

$$1 - p = \frac{np(1-p)}{np} = \frac{Var(T)}{E(T)} = \frac{3.0624}{25.52} = 0.12$$

First find the parameters p and n.

So $p = 1 - 0.12 = 0.88$

$n \times 0.88 = 25.52$

$n = 25.52 \div 0.88$

$= 29$

$$P(T > 27) = P(T = 28) + P(T = 29)$$
$$= \binom{29}{28} \times 0.88^{28} \times 0.12^1 + \binom{29}{29} \times 0.88^{29} \times 0.12^0$$

$T \sim B(29, 0.88)$, so T can take integer values from 0 to 29 inclusive.

$$= 0.097072\ldots + 0.024546\ldots$$
$$= 0.122$$

EXERCISE 7B

1 Calculate the exact expectation and variance of each of the following discrete random variables.

a $V \sim B(6, 0.3)$ **b** $W \sim B(15, 0.42)$

c $X \sim B(180, 0.85)$ **d** $Y \sim B(44, 0.73)$

2 Given that $X \sim B(64, 0.125)$, calculate:

a $E(X)$ and $Var(X)$ **b** $P[X = E(X)]$.

3 Given that $Q \sim \text{B}(7, 0.2)$, calculate:

a $\text{P}(Q \neq 3)$

b $\text{P}[Q < \text{E}(Q)]$.

4 Given that $X \sim \text{B}(n, p)$, $\text{E}(X) = 27$ and $\text{Var}(X) = 14.85$, find:

a the value of n and of p

b $\text{P}(X = 30)$.

5 Given that $G \sim \text{B}(n, p)$, $\text{E}(G) = 49$ and $\text{Var}(G) = 20\frac{5}{12}$, find:

a the parameters of the distribution of G

b $\text{P}(G = 50)$.

6 V has a binomial distribution where $\text{E}(V) = 2\frac{1}{2}$ and $\text{Var}(V) = \frac{5}{12}$. Find the values of n and p, and use these to draw up the probability distribution table for V.

7 Figures show that 32% of people taking their driving test in a certain town fail at their first attempt.

a From a random sample of 50 people in the town, how many are expected to fail the test at their first attempt?

b Find the probability that the expected number of failures occur.

M 8 In a certain district, one thousand one-litre samples of household tap water were tested. It was found that 45 of the samples contained the microorganism *Cryptosporidium parvum*.

a A number n of additional one-litre samples are taken. Based on the evidence of the first 1000 samples, a total of 63 samples are expected to contain the microorganism. Show that $n = 400$.

[You may assume that the number of one-litre samples containing the microorganism is well modelled by a binomial distribution.]

b After taking the first 1000 samples, but before taking the additional samples, a biologist calculated the probability that exactly 63 out of 1400 samples will contain the microorganism. This was found to be:

$$\binom{1400}{63} \times 0.045^{63} \times 0.955^{1337} = 0.0514$$

Explain the error that the biologist has made in this case, and calculate the correct probability.

TIP

If your calculator cannot handle such large numbers of combinations, use spreadsheet software such as Excel with the formula =COMBIN(n, r).

PS 9 The 364 students at a school are divided into 13 equal-sized classes. It is known that exactly 17 of the students are left-handed.

a Use a binomial approximation to find the probability that a randomly selected class contains exactly two left-handed students. Determine whether this is an under- or over-estimate of the true probability.

b Ten classes are selected at random. Find the probability that all but one of them contains exactly two left-handed students.

7.2 The geometric distribution

WORKED EXAMPLE 7.3

Given that $X \sim \text{Geo}(0.83)$, find correct to 4 significant figures the value of:

a $P(X = 4)$ b $P(X \neq 3)$ c $P(X \geqslant 5)$.

Answer

a $P(X = 4) = p(1 - p)^3$
$= 0.83 \times 0.17^3$
$= 0.004078$

$p = 0.83$ and $1 - p = q = 0.17$

b $P(X \neq 3) = 1 - p(1 - p)^2$
$= 1 - (0.83 \times 0.17^2)$
$= 0.9760$

Use $P(X \neq 3) = 1 - P(X = 3)$.

c $P(X \geqslant 5) = P(X > 4)$
$= 0.17^4$
$= 0.0008352$

Use $P(X > r) = (1 - p)^r$ or q^r.

EXERCISE 7C

1 The random variable T follows a geometric distribution and $T \sim \text{Geo}(0.82)$. Find the exact value of:

a $P(T = 3)$ b $P(T \neq 3)$ c $P(T \leqslant 2)$.

2 Given that $X \sim \text{Geo}(0.7)$, find the exact value of:

a $P(X = 5)$ b $P(X < 4)$.

3 The random variable Y follows a geometric distribution and $P(Y = 2) = 0.25$.

a Find the value of the distribution's parameter, p.

b Find the probability that $Y \leqslant 3$.

4 An athlete estimates that he has a 20% chance of coming first and a 25% chance of coming second in each race that he enters. Calculate an estimate of the probability that:

a he comes first in neither of his first two races

b the first race in which he comes either first or second is the third race that he enters.

5 A six-sided die is rolled 1200 times. A 6 is obtained exactly 450 times.

a Does the die appear to be fair? Give a reason for your answer.

b Write down a suitable estimate for the value of p, the probability of obtaining a 6 with each roll of the die, giving your answer in its simplest form.

c Use your answer to part **b** to estimate the probability that:

i exactly six 6s are obtained with ten rolls of the die,

ii the first 6 is obtained on the third roll of the die.

6 In a large survey of adults, it is found that exactly 75% voted at the last election. A sample of 60 of these adults is taken.

a Find the expected number of adults in the sample who did not vote at the last election.

b Show that the variance of the number of adults who voted is 11.25.

c Find the probability that the first person in the sample who did not vote is the fourth person asked.

7 Four-fifths of all children say that they prefer to watch movies at home rather than at the cinema. Children are selected at random, one after the other, and asked to state their preference. Find the probability that:

a the first child who prefers to watch at the cinema is the third child asked

b the first child who prefers to watch at home is not one of the first three asked.

8 $X \sim \text{B}(2, 0.8)$ and $Y \sim \text{Geo}(0.1)$ are independent discrete random variables. Find, correct to 4 significant figures, the probability that:

a $X = 2$ and $Y > 2$

b $XY > 4$

c $\frac{X}{Y} \leqslant 0.5$.

9 Two independent random variables are $S \sim \text{B}(6, 0.3)$ and $T \sim \text{Geo}\left(\frac{1}{4}\right)$. Given that $S > 4$, find the probability that $S - T < 3$.

Mode of the geometric distribution and expectation of the geometric distribution

WORKED EXAMPLE 7.4

The sides of a fair pentagonal spinner are numbered 2, 3, 4, 5 and 6. Let the number of spins up to and including the spin on which the first 6 is obtained be X.

a State the mode and the expectation of X.

b Find, correct to 3 significant figures, the probability that the spinner is spun fewer than nine times.

Answer

a The mode is 1. — The mode of all geometric distributions is 1.

$X \sim \text{Geo}(0.2)$, so $\text{E}(X) = \frac{1}{0.2} = 5$. — Define the distribution and use $\text{E}(X) = \frac{1}{p}$.

b $\text{P}(X < 9) = \text{P}(X \leqslant 8)$
$= 1 - (1 - 0.2)^8$
$= 0.832$
— Use $\text{P}(X \leqslant r) = 1 - (1 - p)^r$ or $1 - q^r$, and give the answer to the required degree of accuracy.

EXERCISE 7D

1 Given that X follows a geometric distribution and that $\text{E}(X) = 1.6$, find $\text{P}(X = 1)$.

2 The random variable Q follows a geometric distribution. Given that $\text{P}(Q = 2) = 0.16$, find the two possible values of:

a the parameter p **b** $\text{E}(Q)$.

3 Independent trials are carried out in which the probability of success is 0.05. Find, correct to 3 significant figures, the probability that the first success occurs before the fifth trial.

4 The sides of a fair heptagonal spinner are numbered 2, 2, 3, 4, 6, 8 and 9. It is spun up to and including the spin on which it first shows an odd number. Find the probability that it is spun more than seven times.

5 An unfair four-sided die is numbered 1, 3, 4 and 8. The score on the die is denoted by X. It is given that:

$\text{P}(X = x) = \frac{x + 2}{k}$ where k is a constant.

a Show that $k = 24$.

The die is rolled up to and including the roll on which an odd number is first obtained.

b What is the most likely number of times that the die will be rolled?

c Find the probability that the die is rolled fewer than three times.

6 Vehicles are tested annually for roadworthiness, and it is known that 40% of vehicles are in good enough condition to pass the test. Of those that fail the test, 25% of the owners spend between \$50 and \$100 on repairs so that their vehicles pass, and the remaining owners spend between \$100 and \$200 so that their vehicles pass.

a Calculate an estimate of the amount that a randomly selected owner can expect to spend on repairs in order for their vehicle to pass the test.

b Find the probability that the fourth randomly selected vehicle is the first one to pass the test without the owner spending money on repairs.

7 A fair four-sided die is numbered 1, 3, 4 and 6. A fair eight-sided die is numbered 1, 1, 1, 3, 3, 4, 5 and 6. The two dice are rolled, simultaneously, up to and including the first occasion on which they both score a 1.

a Determine how many times the dice:

i are most likely to be rolled

ii are expected to be rolled.

b Find the least possible value of n such that the probability that the dice are rolled fewer than n times is greater than $\frac{2}{3}$.

PS 8 On a fairground stall, a prize worth \$25 is offered for successfully throwing a hoop over a pole. Evan's chance of success on each throw, independent of all other throws, is estimated to be 12.5%. He pays the owner of the stall \$1.50 for each throw and continues throwing the hoop until he wins the prize.

a Find the expected number of throws that Evan needs to win the prize.

b Calculate the probability that Evan pays more to win the prize than it is worth.

P 9 Two independent random variables are $X \sim \text{Geo}(p)$ and $Y \sim \text{Geo}\left(\frac{p}{2}\right)$. Find the value of the constants a and b, given that:

$$P(X + Y < 4) = p^2(a + bp)$$

END-OF-CHAPTER REVIEW EXERCISE 7

1 Given that $X \sim B(10, 0.3)$, find:

a $P(X = 2)$

b $P(0 < X \leqslant 2)$.

2 At a fairground stall, people are given three darts to throw at a target. The number of darts that hit the target, D, follows a binomial distribution with $p = 0.85$. Find:

a $E(D)$ and $Var(D)$

b the exact probability that two of a person's three darts hit the target.

3 The probability distribution table for the random variable V is shown.

v	0	1	2
$P(V = v)$	0.6724	0.2952	0.0324

Explain how you know that V follows a binomial distribution.

4 The random variable T follows a geometric distribution with $p = \frac{2}{11}$. Find:

a the mode and the expectation of T

b $P(T = 3)$

c $P(T > 8)$.

5 Given that $Y \sim B(7, 0.2)$, find the smallest value of a for which $P(Y < a) > 0.9$.

6 Analysis revealed that 42% of the cars advertised for sale on a particular website were priced over their market value. Find the probability that, in a random sample of 20 cars on this website, between 8 and 11 inclusive were priced over their market value.

M 7 Priya has noted the different types of bird that come to feed on the bird table in her garden. Over the past week, she has recorded the following numbers.

Sparrows	Pigeons	Weavers	Mousebirds	Waxbills	Bulbuls
205	175	67	64	22	17

One morning she begins her observation and decides to count the number of birds that come to feed, up to and including the first bird that is not a sparrow or a pigeon. The number of birds that she counts is denoted by X.

a What do you need to assume about the behaviour of the birds, if you wish to model the distribution of X by a geometric distribution?

b Find the probability that Priya counts fewer than three birds.

c Calculate $P(X \geqslant 8)$.

8 Two independent random variables, X and Y, are such that $X \sim B(3, 0.6)$ and $Y \sim Geo(0.4)$. One value of X and one value of Y are selected at random.

a List all of the ways in which $X + Y$ can be equal to 3.

b Find $P(X + Y \neq 3)$.

9 At a certain college, the ratio of female students to male students is 11:10. It is known that 25% of the female students and 37% of the male students belong to a sports club. Twenty randomly selected students are interviewed, one at a time. Find the probability that:

a the first student interviewed is a male who does not belong to a sports club

b the second student interviewed is a female who belongs to a sports club

c the second student interviewed is the first female asked who belongs to a sports club

d none of the first five students interviewed is a male who belongs to a sports club.

Chapter 8
The normal distribution

- Sketch normal curves to illustrate distributions or probabilities.
- Use a normal distribution to model a continuous random variable and use normal distribution tables.
- Solve problems concerning a normally distributed variable.
- Recall and recognise conditions under which the normal distribution can be used as an approximation to the binomial distribution, and use this approximation, with a continuity correction, in solving problems.

8.1 Continuous random variables

Representation of a probability distribution and the normal curve

WORKED EXAMPLE 8.1

The normal probability distributions of two related variables, S and T, are represented by their probability density functions (PDFs) in the diagram. Give a brief summary, comparing the mean and standard deviation of the two distributions, explaining each comparison that you make.

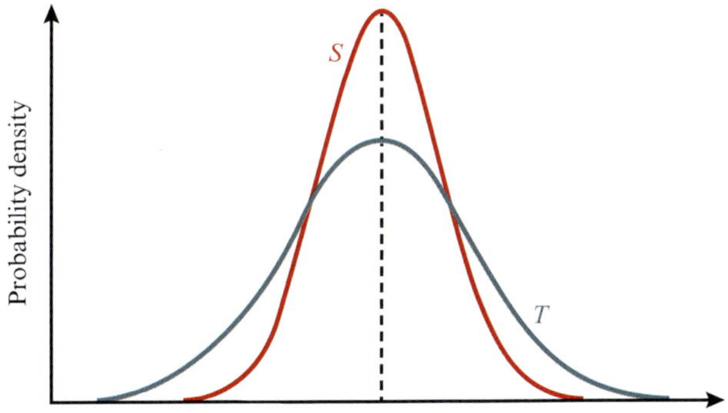

Answer

The curves are centred on the same value, so the mean of S and the mean of T are equal.

The curve for T is wider and shorter than the curve for S, so values of T are more widely spread than values of S. The standard deviation of T is greater than the standard deviation of S.

EXERCISE 8A

1 State whether each of the following describes a continuous random variable, a discrete random variable, or neither:

a the times taken by gymnasts to perform their routines
b the numbers of points awarded to gymnasts for their routines
c the contestant number worn by a gymnast during a particular routine
d the volume of noise made by spectators when applauding gymnasts after their routines.

2 The graph shown illustrates the probability distributions of two variables X and Y. It is given that $X \sim N(15, \sigma^2)$ and $Y \sim N(\mu, 9)$.

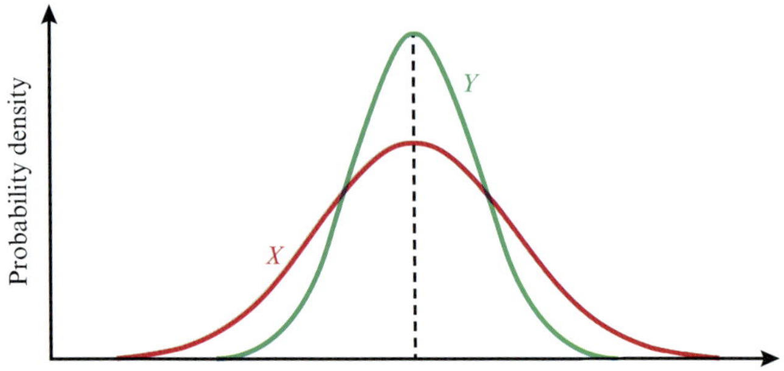

a Use the graphs to write down:

i the value of μ

ii the most accurate possible statement about the value of σ.

b Indicate which of the following pair of probabilities has the greater value, given that the graphs intersect where $X = Y = 11$ and where $X = Y = 19$.

i $P(X > 19)$ or $P(Y < 11)$

ii $P(11 < X < 19)$ or $P(11 < Y < 19)$

3 The continuous random variable T has a normal distribution with mean 56 and standard deviation 12. Write down the following statements, inserting an appropriate mathematical symbol or number into each.

a $P(T < 56)$ …… $P(T > 56)$

b $P(T > 60)$ …… $P(T < 50)$

c $P(20 < T < 92) \approx$ ……

d $P(47 < T < 51)$ …… $P(58 < T < 62)$

> **TIP**
>
> You are advised to sketch a graph to help answer this question.

4 Two brands of tea, A and B, are sold in boxes labelled '200 grams'. The mass of tea in boxes of both brands is normally distributed with mean 204 g for A and 208 g for B. The variance of the mass of tea in the boxes of both brands is $16.5\,g^2$. Describe any similarities and differences between the PDFs representing these two distributions.

M **5** The heights of a large group of women are normally distributed with mean 158 cm and standard deviation 5 cm. The women have, on average, 2.5 children each. The heights of these children are also normally distributed with a mean that is less than 158 cm and a standard deviation that is greater than 5 cm. On a single diagram, sketch a graph representing each of these height distributions.

M 6 The voltages of a large number of newly manufactured car batteries are normally distributed with mean 12.8 and standard deviation 0.5. A symmetrical curve is drawn to represent this distribution.

After each battery has been used to start a car engine, its voltage is checked and a new curve representing the distribution of the battery voltages is drawn. It is discovered that the new curve can be mapped onto the original by the vector $\begin{pmatrix} 0.15 \\ 0 \end{pmatrix}$. What useful information does this provide about the batteries?

M 7 The masses of the pregnant cows on a large farm are denoted by X, where $X \sim N(\mu, \sigma^2)$. Each cow gives birth to a single calf. The masses of the cows and calves together are denoted by Y, which has mean m and standard deviation s.

a Use inequalities to compare the values of:

i μ and m ii σ and s.

b Give brief details of another way in which the distribution of Y differs from the distribution of X.

M 8 A factory manufactures a particular style of shoe in a variety of sizes. The mass, X, of these pairs of shoes is normally distributed with mean 0.764 kg and variance 0.00024 kg^2. Define the distribution of the masses of the individual shoes, Y, stating what assumption you must make in order to do so.

8.2 The normal distribution

The standard normal variable Z

WORKED EXAMPLE 8.2

The body temperature of patients attending a clinic is normally distributed with mean 37.0 °C and standard deviation 0.3 °C. From a random sample of 600 patients, how many are expected to have temperatures outside the range 36.7 to 37.3 °C?

Answer

$36.7 = 37.0 - 0.3 = \mu - \sigma$
$37.3 = 37.0 + 0.3 = \mu + \sigma$

The lower and upper boundaries of the given range are both one standard deviation from the mean.

68.26% of the temperature values are within this range.

Recall that 68.26% of the values in a normal distribution lie within one standard deviation of the mean. The required proportion of patients is represented by the shaded areas in the diagram.

$100 - 68.26 = 31.74\%$
31.74% of $600 = 190.44$

190 patients are expected to have temperatures outside the range 36.7 to 37.3 °C.

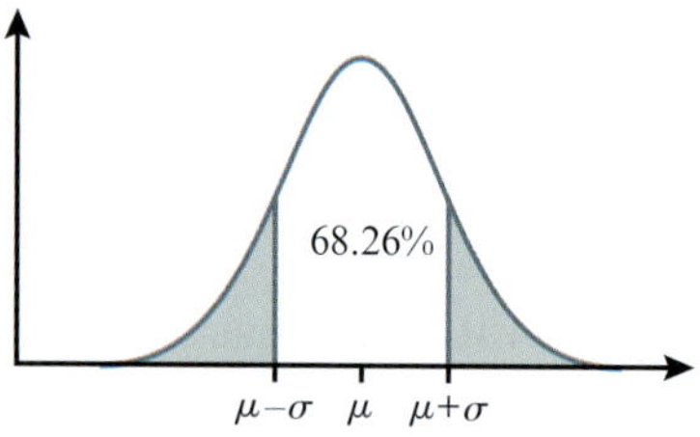

EXERCISE 8B

1 $Z \sim N(0, 1)$ Find the following probabilities.

a $P(Z < 1.23)$
b $P(Z \leqslant 2.468)$
c $P(Z < 0.157)$
d $P(Z \geqslant 1.236)$
e $P(Z > 2.378)$
f $P(Z \geqslant 0.588)$
g $P(Z > -1.83)$
h $P(Z \geqslant -2.057)$
i $P(Z > -0.067)$
j $P(Z \leqslant -1.83)$
k $P(Z < -2.755)$
l $P(Z \leqslant -0.206)$
m $P(Z < 1.645)$
n $P(Z \geqslant 1.645)$
o $P(Z > -1.645)$
p $P(Z \leqslant -1.645)$

2 The random variable Z is distributed such that $Z \sim N(0, 1)$. Find these probabilities.

a $P(1.15 < Z < 1.35)$
b $P(1.111 \leqslant Z \leqslant 2.222)$
c $P(0.387 < Z < 2.418)$
d $P(0 \leqslant Z < 1.55)$
e $P(-1.815 < Z < 2.333)$
f $P(-0.847 < Z \leqslant 2.034)$

3 The random variable Z is distributed such that $Z \sim N(0, 1)$. Find these probabilities.

a $P(-2.505 < Z < 1.089)$
b $P(-0.55 \leqslant Z \leqslant 0)$
c $P(-2.82 < Z < -1.82)$
d $P(-1.749 \leqslant Z \leqslant -0.999)$
e $P(-2.568 < Z < -0.123)$
f $P(-1.96 \leqslant Z < 1.96)$
g $P(-2.326 < Z < 2.326)$
h $P(|Z| \leqslant 1.3)$
i $P(|Z| > 2.4)$

4 The random variable $Z \sim N(0, 1)$. In each part, find the value of s, t, u or v.

a $P(Z < s) = 0.6700$
b $P(Z < t) = 0.8780$
c $P(Z < u) = 0.9842$
d $P(Z < v) = 0.8455$
e $P(Z > s) = 0.4052$
f $P(Z > t) = 0.1194$

g $P(Z > u) = 0.0071$ h $P(Z > v) = 0.2241$ i $P(Z > s) = 0.9977$

j $P(Z > t) = 0.9747$ k $P(Z > u) = 0.8496$ l $P(Z > v) = 0.5$

m $P(Z < s) = 0.0031$ n $P(Z < t) = 0.0142$ o $P(Z < u) = 0.0468$

p $P(Z < v) = 0.4778$ q $P(-s < Z < s) = 0.90$ r $P(-t < Z < t) = 0.80$

s $P(-u < Z < u) = 0.99$ t $P(|Z| < v) = 0.50$

5 The annual income of a person in full-time employment in a certain town is normally distributed with mean \$27 000 and standard deviation \$4500. Given that there are 43 483 people in full-time employment in the town, estimate the number of people whose annual income is between \$18 000 and \$36 000.

6 The random variable X has a normal distribution. What proportion of the values of X exceed the mean by more than 0.7 of a standard deviation?

7 The random variable Y is normally distributed. Find the value of k, given that 2.5% of the values of Y are more than k standard deviations less than the mean.

8 Readings of a continuous random variable are found to be normally distributed. Given that 4262 of the readings lie within one standard deviation of the mean, estimate the number of readings that lie more than one but less than two standard deviations from the mean.

PS 9 Given that $Z \sim N(0, 1)$, find the value of k correct to 3 significant figures, if $P(Z > k) = 2k$.

Standardising a normal distribution

WORKED EXAMPLE 8.3

Given that $Q \sim N(50, 36)$, find $P(Q \leqslant 55.1)$, giving your answer correct to 1 significant figure.

Answer

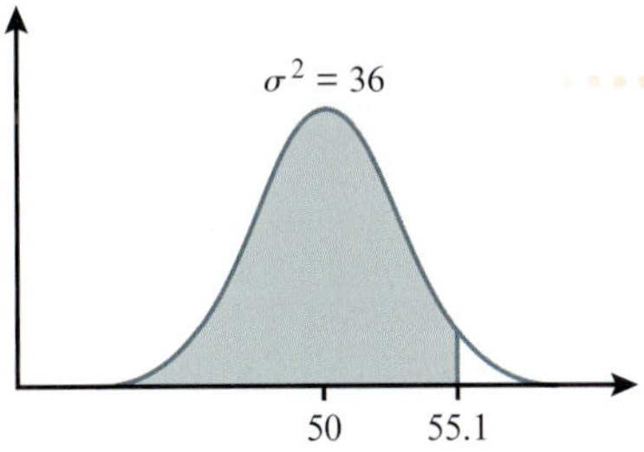

Make a sketch, including the given information. The probability required is represented by the area to the left of the vertical line at 55.1.

$$P(Q \leqslant 55.1) = \Phi\left(\frac{55.1 - 50}{\sqrt{36}}\right)$$
$$= \Phi(0.85)$$
$$= 0.8023$$

$P(Q \leqslant 55.1) = 0.8$

Check the degree of accuracy required in the final answer.

EXERCISE 8C

1 The variable X is normally distributed with mean 20 and standard deviation 4.

a Define the distribution of X in the form $X \sim N(\mu, \sigma^2)$.

b Find the standardised value for $X = 26$.

c Use your answer to part **b** to find:

i $P(X < 26)$ ii $P(X < 14)$.

2 a Given that $Z \sim N(0, 1)$, find $P(Z < 1.555)$.

b Given that $Y \sim N(54, 37)$, find $P(Y < 54)$.

c Given that $X \sim N(40, 81)$, find $P(X < 46.3)$.

d Given that $W \sim N(1200, 2500)$, find $P(W > 1178)$.

e Given that $V \sim N(823.6, 400)$, find $P(V < 800)$.

3 Given that $X \sim N(20, 16)$, find the following probabilities.

a $P(X \leqslant 26)$ b $P(X > 30)$ c $P(X \geqslant 17)$ d $P(X < 13)$

4 Given that $X \sim N(24, 9)$, find the following probabilities.

a $P(X \leqslant 29)$ b $P(X > 31)$ c $P(X \geqslant 22)$ d $P(X < 16)$

5 Given that $X \sim N(50, 16)$, find the following probabilities.

a $P(54 \leqslant X \leqslant 58)$ b $P(40 < X \leqslant 44)$

c $P(47 < X < 57)$ d $P(39 \leqslant X < 53)$

e $P(44 \leqslant X \leqslant 56)$

6 The random variable X has a normal distribution. The mean is μ (where $\mu > 0$) and the variance is $\frac{1}{4}\mu^2$.

a Find $P(X > 1.5\mu)$.

b Find the probability that X is negative.

7 Given that $X \sim N(35.4, 12.5)$, find the values of s, t, u and v correct to 1 decimal place when:

a $P(X < s) = 0.96$ b $P(X > t) = 0.9391$

c $P(X > u) = 0.2924$ d $P(X < v) = 0.1479$.

8 X is a normally distributed random variable with mean μ and standard deviation σ.

a Given that $P(X < 10) = 0.9332$, show that $10 - \mu = 1.5\sigma$.

b Given that $P(X < 15.05) = 0.9940$, form an equation in μ and σ.

c Solve the simultaneous equations from part **a** and part **b** to find the value of μ and of σ.

d Use your answers to part **c** to find $P(X > 8)$.

9 X has a normal distribution, and $P(X > 73.05) = 0.0289$. Given that the variance of the distribution is 18, find the mean.

10 X is distributed normally, $P(X \geqslant 59.1) = 0.0218$ and $P(X \geqslant 29.2) = 0.9345$. Find the mean and standard deviation of the distribution, correct to 3 significant figures.

11 $X \sim N(\mu, \sigma^2)$, $P(X \geqslant 9.81) = 0.1587$ and $P(X \leqslant 8.82) = 0.0116$. Find μ and σ, correct to 3 significant figures.

12 For the variable $T \sim N(\mu, \sigma^2)$ it is given that $\sigma = \frac{1}{4}\mu$ and $P(T > \mu + 1) = 0.2$. Calculate $P(T > \mu - 2)$.

8.3 Modelling with the normal distribution

WORKED EXAMPLE 8.4

A factory produces ball-bearings whose diameters are normally distributed with mean 0.601 cm and variance 1.44×10^{-6} cm. What proportion of the ball-bearings have diameters greater than 0.604 cm?

Answer

$D \sim N(0.601, 1.44 \times 10^{-6})$ — Define the distribution of ball-bearing diameters.

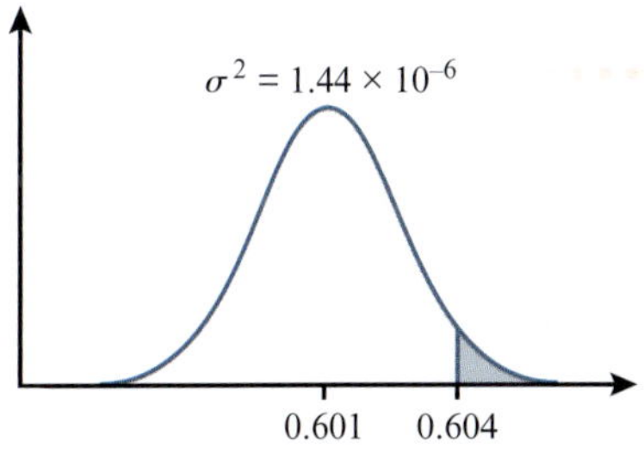

Make a sketch, including the values given. The proportion/probability required is represented by the area to the right of the vertical line at 0.604.

$$P(D > 0.604) = 1 - \Phi\left(\frac{0.604 - 0.601}{\sqrt{0.00000144}}\right)$$
$$= 1 - \Phi(2.5)$$
$$= 1 - 0.9938$$
$$= 0.0062$$

0.62% of the ball-bearings have diameters greater than 0.604 cm. — Clearly state the answer to the question.

EXERCISE 8D

1 The time spent waiting for a medicine to be prepared at a pharmacy is normally distributed with mean 15 minutes and standard deviation 2.8 minutes. Find the probability that the waiting time is:

a more than 20 minutes

b less than 8 minutes

c between 10 minutes and 18 minutes.

2 The lengths of sweetpea flower stems are normally distributed with mean 18.2 cm and standard deviation 2.3 cm.

a Find the probability that the length of a flower stem is between 16 cm and 20 cm.

b 12% of the flower stems are longer than h cm and 20% of the flower stems are shorter than k cm. Find h and k.

c Stem lengths less than 14 cm are unacceptable at a flower shop. In a batch of 500 sweetpea stems, estimate how many would be unacceptable.

3 The life of the Zapower battery has a normal distribution with mean 210 hours. It is found that 4% of these batteries operate for more than 222 hours. Find the variance of the distribution, correct to 2 significant figures.

4 In a statistics examination, 15% of the candidates scored more than 63 marks, while 10% of the candidates scored less than 32 marks. Assuming that the marks were distributed normally, find the mean mark and the standard deviation.

5 Patients recovering from a certain medical procedure spend T hours recovering in hospital, where T is normally distributed with mean 50 and standard deviation 12.

a Find, correct to 3 significant figures, the probability that a randomly selected patient spends 41 hours or more recovering in hospital.

b Find the probability that two randomly selected patients both spend 41 hours or more recovering in hospital.

6 Technical progress reveals that past measurements made by a group of astronomers were subject to normally distributed errors with mean −0.005% and standard deviation 0.81%. Find the probability that this group of astronomers underestimated the distance to a particular star by more than 0.3%.

7 The variable $X \sim \text{N}(16.5, 10)$. Find the probability that more than three out of five randomly selected values of X are between 14.1 and 19.1.

8 A clothing manufacturer makes dresses in four sizes: 8.8% of the dresses are labelled 'small'; 22.5% 'medium'; 36.1% 'large'; the remainder 'extra large'. These four dress sizes equate to a range of body measurements that are normally distributed with mean 86 cm and variance 52 cm^2.

a Find, correct to 2 decimal places, the range of body measurements of customers for whom large is the most suitable size.

b Calculate the probability that a random selection of four dresses contains more than one of any of the four sizes small, medium, large or extra large.

9 The variable Y is normally distributed with mean μ and standard deviation σ. It is given that $P(Y < k) = 0.65$ and that $P(Y < k - 2) = 0.48$. Use this information to find:

a σ b $P(Y > k + 3)$.

PS 10 Omar is employed to lay six floorboards along a corridor, which is 4.2 metres long. He buys six floorboards from stock whose lengths are normally distributed with mean 4.25 m and standard deviation 0.0032 m. He cuts exactly 52 mm off the length of each floorboard. Find the probability that more than one of the six floorboards is too long to fit along the length of the corridor.

8.4 The normal approximation to the binomial distribution

WORKED EXAMPLE 8.5

An ordinary fair die is rolled 306 times. Calculate an estimate of the probability that more than 35 sixes are obtained.

Answer

Let X represent the number of sixes obtained, then $X \sim B\left(306, \frac{1}{6}\right)$.

Name and define the distribution of the discrete random variable.

$np = 51$ and $nq = 255$

Check that the condition for approximating a binomial by a normal distribution (np and nq both > 5) is satisfied.

$\mu = np = 51$, $\sigma^2 = npq = 42.5$

State the parameters μ and σ^2 that are to be used.

X can be approximated by $N(51, 42.5)$.

Define the approximating distribution.

To find $P(X > 35)$, calculate using 35.5. Make a continuity correction because the discrete value $X = 35$ is represented by the class of values $34.5 \leqslant X < 35.5$.

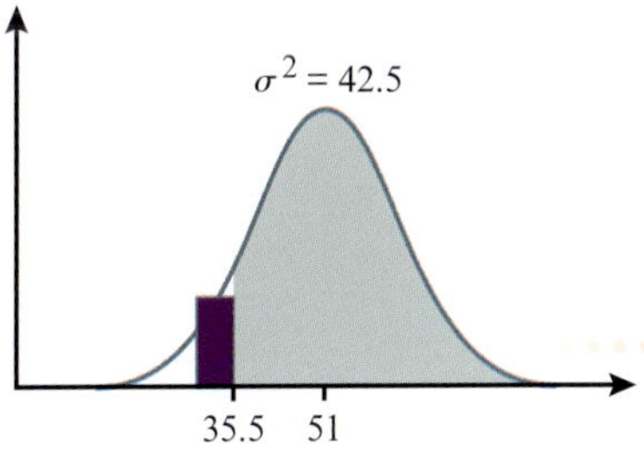

Draw a sketch graph and include a bar to represent the discrete value of 35. Mark the boundary value to be used in the calculations.

$$P(X > 35) \approx P\left(Z > \frac{35.5 - 51}{\sqrt{42.5}}\right)$$
$$= \Phi(2.378)$$
$$= 0.9913$$

The required probability is represented by the area to the right of the vertical line at 35.5.

The probability that more than 35 sixes are obtained is approximately 0.991.

Clearly state the answer to the question.

EXERCISE 8E

1 The probability of tossing a head with a biased coin is 0.62. The coin is tossed 100 times.

a Find the expected number of heads.

b Find the variance of the number of heads.

c A student wishes to use an approximation to estimate the probability that fewer than 70 heads are obtained when the coin is tossed 100 times. To do this, she calculates:

$$\Phi\left(\frac{70-62}{\sqrt{23.56}}\right) = \Phi(1.648) = 0.950$$

i Which of the three values, 70, 62 or 23.56, used by the student is incorrect? Explain why this value is incorrect, and state the correct value that she should use.

ii Calculate the probability of obtaining fewer than 70 heads when the coin is tossed 100 times.

2 The discrete random variable X has a binomial distribution where $n = 85$ and $p = 0.44$.

a Evaluate np and nq.

b What do the two values found in part **a** tell you about the distribution of X?

c Use a suitable approximation to find $P(X \geqslant 35)$.

3 A random variable, X, has a binomial distribution with parameters $n = 40$ and $p = 0.3$. Use a suitable approximation, which you should show is valid, to calculate the following probabilities.

a $P(X \geqslant 18)$

b $P(X < 9)$

c $P(X = 15)$

d $P(11 < X < 15)$

4 In a certain country, 12% of people have green eyes. In a sample of 50 people from this country, find the probability that:

a 12 or more of them have green eyes

b between 3 and 10 (inclusive) of them have green eyes.

Show that your approximation is valid.

5 At an election there are two political parties, X and Y. On past experience, twice as many people vote for party X as for party Y.

a In an opinion poll, a researcher samples 90 voters. Find the probability that 70 voters or more say they will vote for party X at the next election.

b If 2000 researchers each question 90 voters, how many of these researchers would be expected to record '70 or more will vote for party X'?

6 Fuses are packed in a box that contains 20 fuses. 5% of the fuses are faulty. The boxes are packed in crates that contain 50 boxes. Find the probabilities of the following events, clearly stating which distribution you are using and why.

a A box contains two faulty fuses.

b A box contains at least one faulty fuse.

c A crate contains between 35 and 39 (inclusive) boxes with at least one faulty fuse.

7 The sides of a biased four-sided spinner are labelled A, B, C, D. When the spinner was spun 50 times, a B was obtained on exactly 27 occasions.

a Using a suitable approximation and continuity correction, estimate the probability that more than 250 Bs will be obtained with 500 spins.

b The spinner is spun 500 times on three occasions. The discrete random variable B is the number of occasions on which more than 250 Bs are obtained.

i Give the name of the type of distribution that B has, and write down the value of the distribution's two parameters.

ii Hence, find the probability that more than 250 Bs are obtained on exactly two of these three occasions.

8 The rules for Australian football award six points for a 'goal' and one point for a 'behind'. In a particular league, the total number of points scored by the two teams in a game is normally distributed with mean 162.7 and standard deviation 24.3.

a The probability that a total of more than 199 points are scored in a randomly selected game is estimated to be 0.065. Explain why a continuity correction was necessary to make this calculation.

b Ten games are selected at random and the discrete random variable, X, is the number of games in which more than 199 points are scored. Define the distribution of X.

c Find the probability that more than 199 points are scored in exactly one out of ten randomly selected games.

9 An internet service provider has sent out a questionnaire which asked customers whether they were happy, satisfied or displeased with the services provided. The customers' responses are shown in the table.

	Happy	Satisfied	Displeased
No. customers	987	4277	2961

a A random selection of 50 customers is taken. Calculate, to an appropriate degree of accuracy, the probability that:

i exactly half of them said they were satisfied with the service

ii more than 15 said they were displeased with the service

iii not more than four said they were happy with the service.

b Which of the probabilities found in part **a** is the most reliable, and which is the least reliable? Give a reason for each of your answers.

M **10** Precision-made cylinders for use in the construction of jet engines are made by company C. They have an external cross-sectional area of exactly $107.5770\,\text{mm}^2$.

Non-stretch packaging tubes are made by company T. The circumferences of the tubes are normally distributed with mean 37.0 mm and variance $0.0395\,\text{mm}^2$.

To test whether the tubes are suitable for protecting the cylinders during transportation, a worker attempts to slot a cylinder into 200 randomly selected packaging tubes.

a Find the probability that the cylinder will not fit into at least 30 of the packaging tubes.

b Which of the two companies would you advise to make adjustments to their products? Suggest two different options that this company has for making suitable adjustments.

END-OF-CHAPTER REVIEW EXERCISE 8

1 Given that $Z \sim N(0, 1)$, find:

a $P(Z < 1.636)$ b $P(Z > 1.915)$.

2 The random variable X follows a normal distribution with mean μ and standard deviation σ. Find, correct to 4 significant figures, the probability that X takes a value in the range $\mu \leqslant X < \mu + \frac{3}{5}\sigma$.

3 A continuous random variable X has a normal distribution with mean 15 and standard deviation σ. Given that $P(X < 21.74) = 0.75$, find:

a σ b $P(X < 23.1)$.

4 The daily amount of refuse, R kilograms, produced by the restaurants in a particular street follows a normal distribution with mean μ and variance σ^2. It is given that $P(R < 120) = 0.8$ and $P(R < 115) = 0.525$.

a Form a pair of simultaneous equations in μ and σ to represent this information.

b Solve your equations to find the values of μ and σ.

c Find the probability that the restaurants in this street produce more than 100 kg of refuse on any particular day.

5 The variable Q is normally distributed. Given that $7\sigma = 2\mu$ and that $P(Q < 30) = 0.063$, find:

a the values of μ and σ b $P(Q > 60)$.

6 The value of houses in a certain district is normally distributed with mean \$235 420 and standard deviation \$28 724.

a Find the percentage of houses in this district with values from \$200 000 to \$275 000.

b Three houses in this district are selected at random. Find the probability that exactly one of them has a value of more than \$275 000.

7 A survey in a city reveals that 82% of children prefer to play indoors rather than outdoors. Find the probability that, in a random sample of 150 children in this city, more than 130 of them prefer to play indoors.

8 A biased six-sided die is such that the probability of rolling three consecutive sixes is equal to 0.064. The die is rolled 500 times. Calculate an estimate of the probability that fewer than 195 sixes are rolled.

9 The numbers of votes cast for the three candidates at an election are represented in a pie chart with sector angles 151.2°, 136.8° and 72°. A random sample of 1200 voters is taken. Given that fewer than 525 voted for the winning candidate, estimate the probability that more than 495 voted for the winning candidate.

P 10 Given that $X \sim N(\mu, \sigma^2)$, show that:
$P[(\mu - \sigma < X < \mu + \sigma) \mid (\mu - 2\sigma < X < \mu + 2\sigma)] \approx \frac{143}{200}$

THE STANDARD NORMAL DISTRIBUTION FUNCTION

If Z is normally distributed with mean 0 and variance 1, the table gives the value of $\Phi(z)$ for each value of z, where

$$\Phi(z) = \mathrm{P}(Z \leqslant z).$$

Use $\Phi(-z) = 1 - \Phi(z)$ for negative values of z.

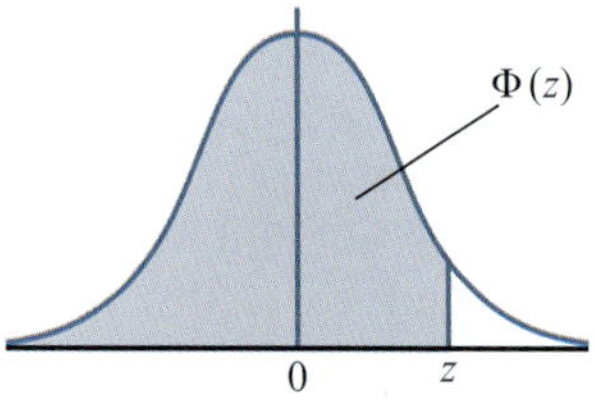

z	0	1	2	3	4	5	6	7	8	9	1	2	3	4	5	6	7	8	9
															ADD				
0.0	0.5000	0.5040	0.5080	0.5120	0.5160	0.5199	0.5239	0.5279	0.5319	0.5359	4	8	12	16	20	24	28	32	36
0.1	0.5398	0.5438	0.5478	0.5517	0.5557	0.5596	0.5636	0.5675	0.5714	0.5753	4	8	12	16	20	24	28	32	36
0.2	0.5793	0.5832	0.5871	0.5910	0.5948	0.5987	0.6026	0.6064	0.6103	0.6141	4	8	12	15	19	23	27	31	35
0.3	0.6179	0.6217	0.6255	0.6293	0.6331	0.6368	0.6406	0.6443	0.6480	0.6517	4	7	11	15	19	22	26	30	34
0.4	0.6554	0.6591	0.6628	0.6664	0.6700	0.6736	0.6772	0.6808	0.6844	0.6879	4	7	11	14	18	22	25	29	32
0.5	0.6915	0.6950	0.6985	0.7019	0.7054	0.7088	0.7123	0.7157	0.7190	0.7224	3	7	10	14	17	20	24	27	31
0.6	0.7257	0.7291	0.7324	0.7357	0.7389	0.7422	0.7454	0.7486	0.7517	0.7549	3	7	10	13	16	19	23	26	29
0.7	0.7580	0.7611	0.7642	0.7673	0.7704	0.7734	0.7764	0.7794	0.7823	0.7852	3	6	9	12	15	18	21	24	27
0.8	0.7881	0.7910	0.7939	0.7967	0.7995	0.8023	0.8051	0.8078	0.8106	0.8133	3	5	8	11	14	16	19	22	25
0.9	0.8159	0.8186	0.8212	0.8238	0.8264	0.8289	0.8315	0.8340	0.8365	0.8389	3	5	8	10	13	15	18	20	23
1.0	0.8413	0.8438	0.8461	0.8485	0.8508	0.8531	0.8554	0.8577	0.8599	0.8621	2	5	7	9	12	14	16	19	21
1.1	0.8643	0.8665	0.8686	0.8708	0.8729	0.8749	0.8770	0.8790	0.8810	0.8830	2	4	6	8	10	12	14	16	18
1.2	0.8849	0.8869	0.8888	0.8907	0.8925	0.8944	0.8962	0.8980	0.8997	0.9015	2	4	6	7	9	11	13	15	17
1.3	0.9032	0.9049	0.9066	0.9082	0.9099	0.9115	0.9131	0.9147	0.9162	0.9177	2	3	5	6	8	10	11	13	14
1.4	0.9192	0.9207	0.9222	0.9236	0.9251	0.9265	0.9279	0.9292	0.9306	0.9319	1	3	4	6	7	8	10	11	13
1.5	0.9332	0.9345	0.9357	0.9370	0.9382	0.9394	0.9406	0.9418	0.9429	0.9441	1	2	4	5	6	7	8	10	11
1.6	0.9452	0.9463	0.9474	0.9484	0.9495	0.9505	0.9515	0.9525	0.9535	0.9545	1	2	3	4	5	6	7	8	9
1.7	0.9554	0.9564	0.9573	0.9582	0.9591	0.9599	0.9608	0.9616	0.9625	0.9633	1	2	3	4	4	5	6	7	8
1.8	0.9641	0.9649	0.9656	0.9664	0.9671	0.9678	0.9686	0.9693	0.9699	0.9706	1	1	2	3	4	4	5	6	6
1.9	0.9713	0.9719	0.9726	0.9732	0.9738	0.9744	0.9750	0.9756	0.9761	0.9767	1	1	2	2	3	4	4	5	5
2.0	0.9772	0.9778	0.9783	0.9788	0.9793	0.9798	0.9803	0.9808	0.9812	0.9817	0	1	1	2	2	3	3	4	4
2.1	0.9821	0.9826	0.9830	0.9834	0.9838	0.9842	0.9846	0.9850	0.9854	0.9857	0	1	1	2	2	2	3	3	4
2.2	0.9861	0.9864	0.9868	0.9871	0.9875	0.9878	0.9881	0.9884	0.9887	0.9890	0	1	1	1	2	2	2	3	3
2.3	0.9893	0.9896	0.9898	0.9901	0.9904	0.9906	0.9909	0.9911	0.9913	0.9916	0	1	1	1	1	2	2	2	2
2.4	0.9918	0.9920	0.9922	0.9925	0.9927	0.9929	0.9931	0.9932	0.9934	0.9936	0	0	1	1	1	1	1	2	2
2.5	0.9938	0.9940	0.9941	0.9943	0.9945	0.9946	0.9948	0.9949	0.9951	0.9952	0	0	0	1	1	1	1	1	1
2.6	0.9953	0.9955	0.9956	0.9957	0.9959	0.9960	0.9961	0.9962	0.9963	0.9964	0	0	0	0	1	1	1	1	1
2.7	0.9965	0.9966	0.9967	0.9968	0.9969	0.9970	0.9971	0.9972	0.9973	0.9974	0	0	0	0	0	1	1	1	1
2.8	0.9974	0.9975	0.9976	0.9977	0.9977	0.9978	0.9979	0.9979	0.9980	0.9981	0	0	0	0	0	0	0	1	1
2.9	0.9981	0.9982	0.9982	0.9983	0.9984	0.9984	0.9985	0.9985	0.9986	0.9986	0	0	0	0	0	0	0	0	0

Critical values for the normal distribution

The table gives the value of z such that $P(Z \leqslant z) = p$, where $Z \sim N(0, 1)$.

p	0.75	0.90	0.95	0.975	0.99	0.995	0.9975	0.999	0.9995
z	0.674	1.282	1.645	1.960	2.326	2.576	2.807	3.090	3.291

Answers

Answers to proof-style questions are not included.

1 Representation of data

Exercise 1A

1 AY; BZ; CX

2 a 255 and 265 b 37.6 g

3 $66\frac{1}{11}$ or 66.1 km/h and 81 km/h

4 a

0	1 2 2 5 6 7
1	1 3 5 6
2	0

Key: **0**| 1 represents 1 item

b 13

5 a 12.8 m b 162

6 a i Quantitative

ii Discrete

b

2	5 8 9
3	4 4 5 7
4	1 2 3 3 5

Key: **2**| 5 represents 25 grapes

7 a

0	4 6	(2)
1	2 5 8	(3)
2	1 5 5 5 7 8 9	(7)
3	0 2 4 6 7	(5)
4	1 3	(2)
5	2	(1)

Key : 2 | 7 means 27 km/h

b

0	3 4 7 8 9 9 9	(7)
1	0 1 2 6 8	(5)
2	1 1 3 7	(4)
3		(0)
4	2	(1)

Key: 2 | 3 means 2.3 hours

8 a

132	2 9	(2)
133	2 6 8 9	(4)
134	1 1 2 2 2 4 5 6 7 7 7 7 8 9	(13)
135	0 1 1 3 3 3 4 4 6	(9)
136	2	(1)
137	0	(1)

Key: 134 | 7 means 1.347 kg

b There would be only one stem (13), which would have all the leaves.

9 3 and 10 rows

10 a

Yellow-rumped		Red-fronted
3 2	**12**	
5 1 0	**13**	2 3 6 9
8 7 2 2	**14**	0 1 5 7 7 7
7 5 3 1	**15**	2 3 5 6 8 9
9 9 8 7 6 5 5 2	**16**	0 5 8 8
9 8 7 6 4 0	**17**	3 4 5
2 1 1	**18**	0 2 6
	19	3 4
	20	0 2

Key: 2|**14**|0 represents 14.2 g for a Yellow-rumped and 14.0 g for a Red-fronted Tinkerbird

b 3

c i 2, 21 and 7.

ii If the mass recorded as 17.5 g is slightly less than 17.5 g, it will be rounded to 15 g.

iii

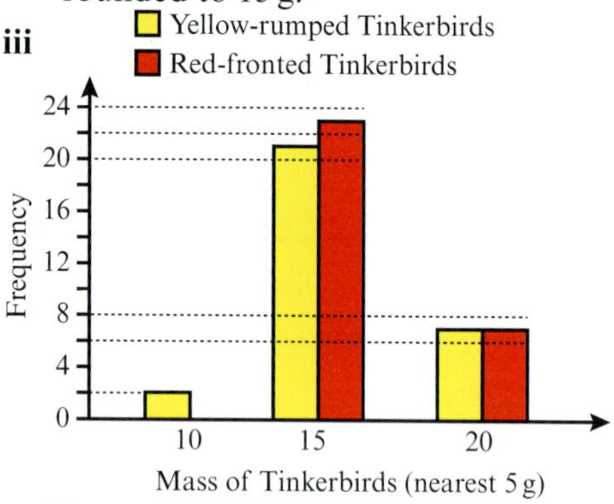

Exercise 1B

1 Boundaries at 45, 60, 75, 90, 105, 120, (e.g. 135)
Densities ∝ 0.8, 2.1, 3.7, 4.8, 1.3, (0.5)

2 Boundaries at 4.5, 9.5, 14.5, 19.5, 24.5, 29.5, 34.5, 44.5
Densities ∝ 0.4, 1.0, 1.6, 2.8, 3.4, 2.2, 0.3

3 a 0, 2.5 and 2.5, 5.5

b Boundaries at 0, 2.5, 5.5, 8.5, 11.5, 15.5
Densities ∝ 6.8, 2, 1.33, 0.67, 0.25

4 **a** and **b** Boundaries at 8.95, 9.95, 10.95, 11.95, 12.95, 13.95, 14.95, 15.95, 16.95
Densities ∝ 1, 4, 7, 6, 9, 8, 7, 3

5 **a** and **b** Boundaries at 16, 20, 30, 40, 50, 60, (e.g. 80)
Densities ∝ 3, 4, 4.4, 4.7, 3.2, (1.25)

6 a i 0.7 up to 1.6 kg ii 1.9 up to 2.3 kg

b Boundaries at 0.7, 1.2, 1.6, 1.9, 2.1, 2.3
Densities ∝ 48, 140, 250, 180, 45

c 27

7 $a = 12$, $b = 3$, $c = 15$, $d = 5$

8 a 399

b Fourth column cannot represent a frequency of 31.5.

9 a $p = 6.1$, $q = 6.2$, $r = 98$

b 7.16 to 7.55 mmol/litre

Exercise 1C

1 a Plot at (0, 0), (16, 14.3), (40, 47.4), (65, 82.7), (80, 94.6), (110, 100).

b 10 (.4) million

2 Plot at (0, 0), (2, 15), (3, 42), (4, 106), (5, 178), (6, 264), (7, 334), (8, 350), (10, 360); about 58 poor days and 14 good days.

3 Assuming x is correct to the nearest km, plot at (0, 0), (4.5, 12), (9.5, 41), (14.5, 104), (19.5, 117), (24.5, 129), (34.5, 132).

a 8 km **b** 14 km

4 **a** Girls: plot at (50, 0), (54, 10), (58, 28), (62, 71), (66, 80).

b Boys: plot at (50, 0), (54, 32), (58, 52), (62, 63), (66, 80).

c **i** 49 or 50 **ii** 22 or 23

5 **a** 11 **b** $n = 119$

6 **a** (3400, 1) and (3625, 12)

b 31 weeks

7 **a** A: plot at (24.5, 0), (29.5, 1), (34.5, 3), (39.5, 6), (44.5, 10), (49.5, 15), (54.5, 21).
B: plot at (24.5, 0), (29.5, 6), (34.5, 11), (39.5, 15), (44.5, 18), (49.5, 20), (54.5, 21).
or
A: plot at (25, 0), (30, 1), (35, 3), (40, 6), (45, 10), (50, 15), (55, 21).
B: plot at (25, 0), (30, 6), (35, 11), (40, 15), (45, 18), (50, 20), (55, 21).

b

No. customers	77	78	79	80	81
No. days (f)	3	4	8	5	1

8 **a** $\frac{x_2 - x_1}{a} = \frac{x_3 - x_2}{b - a} = \frac{x_4 - x_3}{c - b} = \frac{x_5 - x_4}{d - c}$ or $\frac{\text{class frequency}}{\text{class width}}$ is constant

b **i** $x_1 > 0$ **ii** $x_1 \leqslant 0$

Exercise 1D

1 A 1900, B 6536, C 6992, D 4408, E 7524

2 **a** They impart no immediately useful information.

b Numbers or percentages of students who prefer each flavour.

3 **a** 32

b Missing frequencies are 5, 12, 33 and 4.

c Stem-and-leaf diagram will show actual numbers of people employed, but one row will have 33 leaves.
Histogram will give a summary in six equal-width intervals, with frequencies used for column heights, but will not show actual numbers of people employed.

d 2.4 cm and 1.8 cm

4 **a** 165 children visited the library 10 or more but fewer than 21 times.

b e.g. pie chart (angles ≈ 101°, 216°, 43°) with appropriate title and labels

5 **a**

0	6 0 8 0	(4)
1	8 9 2 7 4 1 1 6	(8)
2	7 8 5 6 0 1 9	(7)
3	8 1 7 3 6 4	(6)
4	5 3 2 2	(4)
5	7 5	(2)
6	6 3 2	(3)
7	2 5	(2)
8	5 4 6 2	(4)

Key: 4 | 3 means 43 runs

b The diagram indicates how the scores are distributed. It does not indicate the order in which the scores occurred.

6 Plot cumulative frequency graph at (100, 0), (110, 2), (120, 12), (130, 34), (140, 63), (150, 85), (160, 97), (180, 100); 123 cm.

7 **a** Plot at (10, 16), (20, 47), (30, 549), (40, 1191), (50, 2066), (60, 2349), (80, 2394), (100, 2406), (140, 2410).

b About 74%

c End boundaries are unknown. Use (say) 4 – 10, 100 – 140.

8 **a** Street 1: Plot at (61, 0), (65, 4), (67, 15), (69, 33), (71, 56), (73, 72), (75, 81), (77, 86), (79, 90), (83, 92).
Street 2: Plot at (61, 0), (65, 2), (67, 5), (69, 12), (71, 24), (73, 51), (75, 67), (77, 77), (79, 85), (83, 92).

b 69.8 dB on Street 1, 72.3 dB on Street 2

c Street 1 appears less noisy, in general, than Street 2. For example, there are 56 readings under 71 dB for Street 1, but 24 for Street 2.

9 **a** There is little difference between using a bar chart and a histogram; neither adds much to the frequency table, except that the histogram shows the widths correctly as 5, whereas the bar chart shows them as 4. The bar chart highlights that the data are discrete and shows frequencies directly.

Individual scores presented in a bar chart may be helpful for visual purposes.

b You could no longer use frequency on the bar chart (and certainly not frequency density), so you would be restricted to using a histogram or cumulative frequency graph.

End-of-chapter review exercise 1

1 **a** 502, 503, 504, 505, 506 and 507

b 5.0195 grams

2 **a** 36–55 years

b 28, 46 and 58.5 years

3

	Quantitative	Qualitative	Discrete	Continuous
Distance	✓			✓
Colour		✓		
No. wheels	✓		✓	
Value	✓		✓	

4 **a** Boundaries at 14.5, 25.5, 34.5, 40.5 and 60.5

Densities ∝ 6, 12, 9 and 2.1

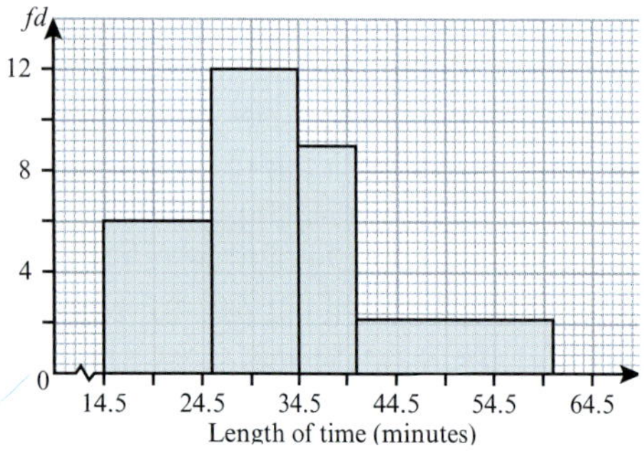

b **i** 120 **ii** 32 or 33

5 **a**

No. bricks (x)	No. houses (f)
6300 – 6500	14
6500 – 7000	18
7000 – 7500	24
7500 – 8000	12
8000 – 9000	7

b Boundaries at 6300, 6500, 7000, 7500, 8000 and 9000

Densities ∝ 0.07, 0.036, 0.048, 0.024 and 0.007

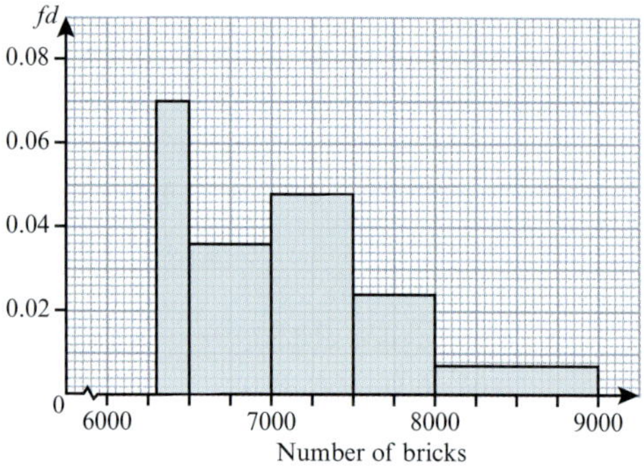

c 7875

6 **a** 54.4%

b Boundaries at −3000, −1200, −500, 0, 500, 1500 and 3000

Densities ∝ 0.02, 0.4, 0.6, 0.72, 0.68 and 0.33

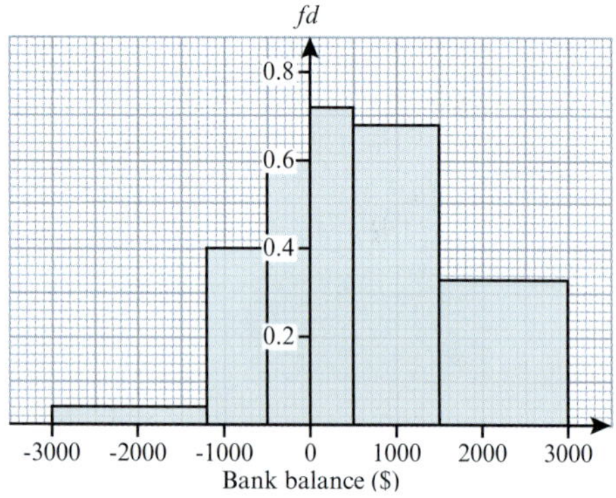

c 0.483 or $\frac{29}{60}$

7 **a** **i** 8 **ii** 12

b Plot at (6, 0), (7, 2), (10, 8), (12, 30), (15, 74), (18, 80) for protein (x).

Plot at (6, 0), (7, 28), (10, 64), (12, 72), (15, 76), (18, 80) for dietary fibre (y).

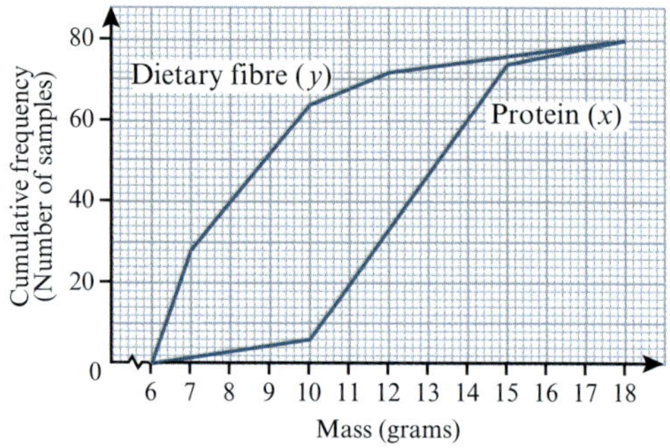

c Yes, more protein than dietary fibre. In these samples, the minimum possible mass of protein is 892 g and the upper boundary of the mass of dietary fibre is 784 g.

8 a Plot at (55, 0), (85, 20), (125, 156), (175, 228), (205, 240).

b Polygon gives ≈ 120; curve gives ≈ 134.

c Polygon gives ≈ 37; curve gives ≈ 35.

2 Measures of central tendency

Exercise 2A

1 a 0 b No mode

c 2 – 3 d Brown

2 0.64, or any equivalent from the list

3 At least two of the five contestants gave 7 correct answers.

4 Frequency densities are $\frac{n}{4} > \frac{n}{5} > \frac{n}{6}$, or 5 – 8 is the narrowest interval.

5 $p = 49$

6 1 hour

7 2 and 26

8 $k = 12$; modal value of $x = 13$

Exercise 2B

1 a 230 b $7\frac{2}{3}$

2 a 10

b Decreased by 100

3 a $x + 19$ b −1 and 77

4 22

5 a 9, 11, 13 and 17 b 11.48 minutes

c 689 seconds or 11 min 29 s

6 163

7 23

8 \$85

9 103.6 litres or 103 600 ml

10 a $x = -\frac{b}{2a}$

b Equation of the axis of symmetry of the graph $y = ax^2 + bx + c$

11 a 167 cm^2 b 20.2 cm

12 a 13.0 cm b 13

13 3.41 and 3.51

Exercise 2C

1 3560

2 $\bar{y} = 13.5$

3 a 13.2

b They all contain 4 green sweets.

4 30.6 °C

5 12

6 53 kg

7 a $\Sigma(x - 2.5)$ b \$12.50

8 630

9 a $37.9 - 29.4 = 8.5$ b 34.5 seconds

Exercise 2D

1 $k = -8$

2 $n = 7$

3 Proof

4 a €24 503.12

b The rate £1 to €1.14446 … found from the two given rates is valid.

5 a 19 b 3.5

6 2.27 minutes

7 a (4, −1) b $(2\bar{y}, 2\bar{x})$

8 a $\frac{10(\bar{x} - 5)}{11}$ b $n = 32$

Exercise 2E

1 5.4 kg; 5.7 kg

2 34

3 a $\frac{n-1}{2}$ and $\frac{n+3}{2}$

b i e.g. 1, 2, 5, 8, 15

ii e.g. 1, 2, 3, 8, 15

iii e.g. 1, 2, 5, 6, 15

4 a Mean 4.875, median 5, mode 6. The data set is too small for the mode to give a reliable estimate of central tendency. The median gives a better idea of a 'typical' mark.

b It has a 'tail' of low values/It is negatively skewed.

5 a $x = 8$ b 14

c Each number occurs once.

6 Mode = 5, mean = $6\frac{5}{11}$ or 6.45, median = 7

7 a $b = 42$, $c = 48$

b i $a = 32$ ii $a = 18$

c 3.0 and 5.0

8 $3.330 - 2.805 = 0.525$ kg

9 a Both 22.0 – 23.9

b Not supported since modal classes the same. Either mean_1 (approx. 23.2 °C) is greater than mean_2 (approx. 22.4 °C), or median_1 (≈ 23 °C) is greater than median_2 (≈ 22 °C).

10 a 196 or 197 seconds

b Proof

c Mean increases, but no effect on median

End-of-chapter review exercise 2

1 a Median = 29.5, mode = 29

b 30.8 c 29.5

2 5.4

3 −0.429

4 $5\frac{2}{3}\,\text{m s}^{-1}$

5 8:9

6 a 26 b 32

7 9

8 a 4 b 16 c 1

9 $n = 40$

3 Measures of variation

Exercise 3A

1 a 17, 9.5 b 9.1, 2.8

2 3

3 a $x = 7$ or 60

b IQR = 33 or 35

4 a Q_1 is at the 25th value.

b $24 \leqslant x < 30$ cm

c 8 cm and 22 cm

5 Monday: $Q_1 = 105$, $Q_2 = 170$, $Q_3 = 258$
Wednesday: $Q_1 = 240$, $Q_2 = 305$, $Q_3 = 377$
Wednesday has greater audiences in general, with less variation.

6 Range = 28, median = 37, IQR = 11

7 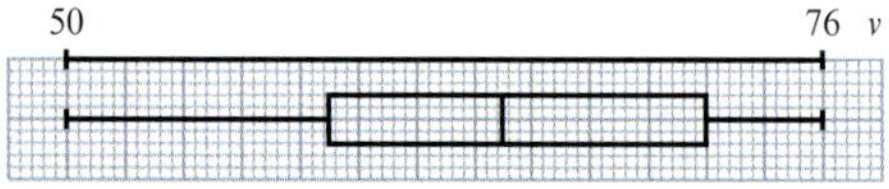

8 a

2	3 6 7 8
3	0 1 2 5 6 8
4	3 4 9 9
5	6 9
6	4 8
7	2

Key: **2** | 3 represents a length of 2.3 cm

b Median = 3.8, IQR = 2.6

c 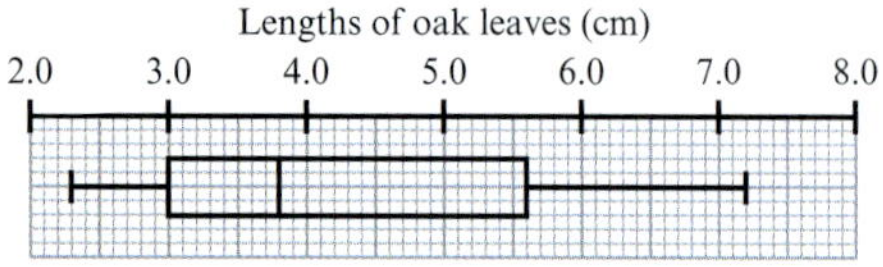

9 39 units

10 a 3.25 b 5.33

11 a 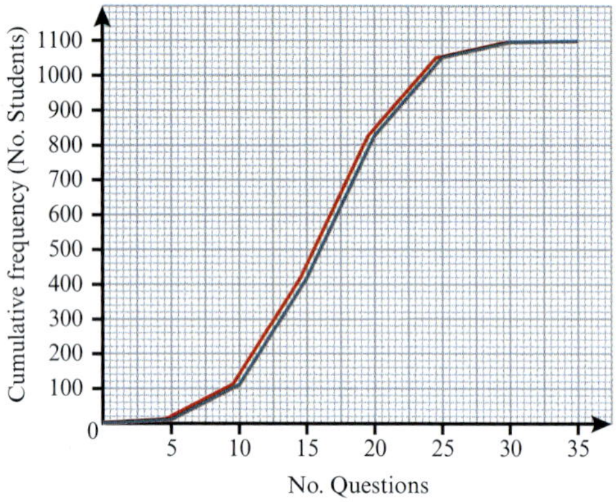

b Median = 16 or 16.6, IQR = 7.2 or 7.3

c 26 or 27

12 a i 29 ii 42 iii 21

b The scores of the men and women are quite similar as the medians are close; men's scores are more spread out.

Exercise 3B

1 a When variance is 0 or 1

b When 0 < variance < 1

2 a 3.49 b 4.28

3 a 75 kg; the five values used in the calculation are not exact.

b 5.30 kg

4 Proof

5 a $r = 9$ b 3.51

6 825 m and 427 m

7 a 1.89, 2.05 and 2.41 m

b i 2.03 m ii $0.0352\,\text{m}^2$

8 a Mean = 251 g, SD = 3.51 g

b Increase the number of classes; weigh more accurately; use more packets.

9 a Mean = 37.5 years, SD = 11.9 years

b In the second company the general age is lower and with smaller spread.

10 a Median = 8.5 minutes, IQR = 9.25 minutes

b

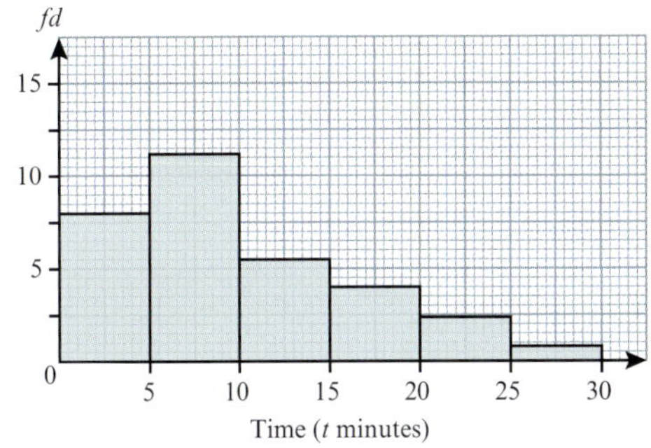

c 10 minutes

d **i** True **ii** True

iii False **iv** True

11 **a** Median = 1.85 m

Quartiles = 1.81 m, 1.89 m

b Negative skew

c Mean = 1.850 m, SD = 0.069 m

12 **a** **i** Mean = 26.9 years, SD = 13.0 years

ii 22.4 years

b Median preferred since distribution skewed; more information given by median.

13 **a** $8\pi\sqrt{6}$ cm **b** 22.5 cm

Exercise 3C

1 15.6

2 50.7 (g), 10.1 (g^2); grams^2

3 ± 1.2

4 4480

5 Females: minimum = 48 kg, Q_1 = 56.5 kg, median = 63 kg, Q_3 = 66.5 kg, maximum = 79 kg

Males: minimum = 60 kg, Q_1 = 67 kg, median = 78 kg, Q_3 = 82 kg, maximum = 87 kg

$\bar{f}$ = 62.3 kg, SD (female) = 7.34 kg,

$\bar{m}$ = 75.6 kg, SD (male) = 8.84 kg

Both distributions have negative skew. Females are lighter than males by about 13 kg on average and less variable than males.

6 **a** Both 15.3

b No; 2.01 and 2.00

7 **a** 300 **b** 60

8 **a** Mean = 268; SD = 252 years

b It is an estimate, correct to within 5 years.

9 32

10 349 545

11 $108.5 \leqslant m < 109.5$ cm; proof

12 **a** **i** −1.6 **ii** 146

b When $\Sigma x = 20$ and $\Sigma y = -20$

Exercise 3D

1 $\bar{x} = 9$, variance = 4

2 0.599 months

3 Mean = 367.5 g; SD = 15.3 g

4 **a** 11

b $a = 7$; $\Sigma x^2 = 1550$

5 **a** 407 grams, 31 grams

b 408 grams, 20 grams

c With the new variety, weight gains are increased on average, but are more variable.

6 14 920

7 **a** $\Sigma(x - \bar{x}) = 0$, $\Sigma(x - \bar{x})^2 = 258$

b Mean = 10, variance = 8.6

8 150

9 **a** 7

b Mean = 5.12, SD = 1.11

10 **a** 21 km

b **i** 23 km

ii 2.5 km; $SD(x + 2) = SD(x)$; Variation is not affected by addition of 2 km to each day's distance.

11 **a** $\bar{x} = 1.92$, $\bar{y} = 6.88$

b $\Sigma y^2 = 1434$ **c** 25.2744

12 **a** 3393 + 2739 = 6132

b Boys = 136 850, girls = 177 400

c 2.75

Exercise 3E

1 399.8 g; that none of the watermelons have been bought/eaten/removed.

2 200%

3 \$11.55

4 **a** 6400 cm^6

b Mean = 18 000 cm^3, variance = 5184 cm^6

5 $p = -3$

6 135

7 18 432

8 1 236 020

9 **a** Mean = 0.025 g, variance = $2.25 \times 10^{-6} g^2$

b Zero (144 g each)

10 **a** 132% of their investment plus \$500

b Mean = \$28 484, SD = \$3748.80

11 **a** $\bar{d} = 2\bar{r}$

b $\text{Var}(c) = 4\pi^2 s^2$

c $\pi(\bar{r}^2 + s^2)$

12 **a** $F = 1.8K - 459.67$

b **i** $\bar{K} = 307; \bar{C} = 33.85; \bar{F} = 92.93$

ii $\Sigma K^2 = 942\,500; \Sigma C^2 = 11\,468.225;$ $\Sigma F^2 = 86\,392.249$

End-of-chapter review exercise 3

1 **a** 196 **b** 3.02 minutes

c **i** 151

ii 148

d 8.4 seconds

2

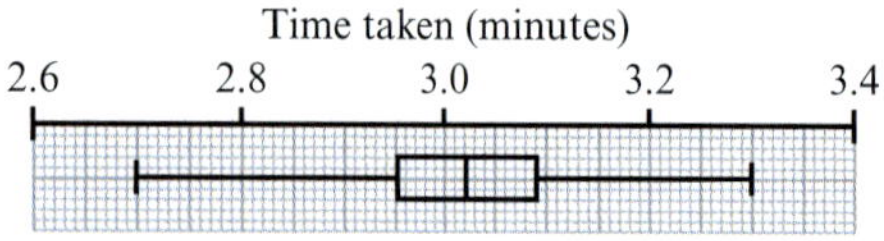

3 Robins: mean = 22.9 cm; SD = 1.29 cm. Sparrows: mean = 22.9 cm, SD = 1.13 cm. The mean wingspans are the same, but the robins' wingspans are more varied.

4 Mean = $\frac{71}{96}$, variance = 1.28

5 27.8 minutes

6 65.5

7 1.89

8 **a** 447.7 **b** 11

9 **a** \$7.80

b Proof; $\Sigma w^2 = 718.66$

c \$4.29

4 Probability

Exercise 4A

1 **a** $\frac{1}{2}$ **b** $\frac{5}{6}$ **c** $\frac{1}{3}$

2 **a** $\frac{2}{9}$ **b** $\frac{4}{9}$

3 **a** $\frac{3}{4}$ **b** $\frac{3}{5}$

4 210

5 **a** A 6 is obtained.

b Neither Frank nor Sabrina is selected.

c At least one goal is scored in the first half.

6 $\frac{1}{64}$

7 **a** 17

b **i** $\frac{1}{13}$ **ii** $\frac{16}{17}$

8 $n = 54°$

Exercise 4B

1 **a** $\frac{1}{30}$ **b** $\frac{29}{30}$

c $\frac{13}{30}$ **d** $\frac{2}{5}$

2 **a** 0.63 **b** 0.45 **c** 0.62

3 $\frac{5}{7}$

4 **a** **i** $\frac{11}{16}$ **ii** $\frac{11}{16}$

b 'Short or female' describes the same students as 'not a tall male'.

5 **a** $\frac{3}{25}$ **b** $\frac{2}{25}$

c $\frac{37}{50}$ **d** $\frac{12}{25}$

6 $\frac{3}{4}$

7 Yes; $P(M) + P(N) = P(M \cup N) = 0.8$

8 0.67

Exercise 4C

1 **a**

2nd roll				
4	5	6	7	8
3	4	5	6	7
2	3	4	5	6
1	2	3	4	5
	1	2	3	4 1st roll

b **i** $\frac{1}{4}$ **ii** $\frac{5}{8}$ **iii** $\frac{3}{16}$

2 **a**

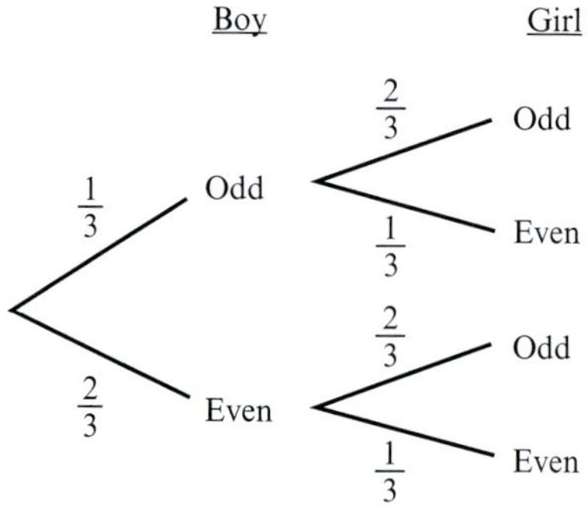

b **i** $\frac{2}{9}$ **ii** $\frac{4}{9}$

3 **a** 0.4

b

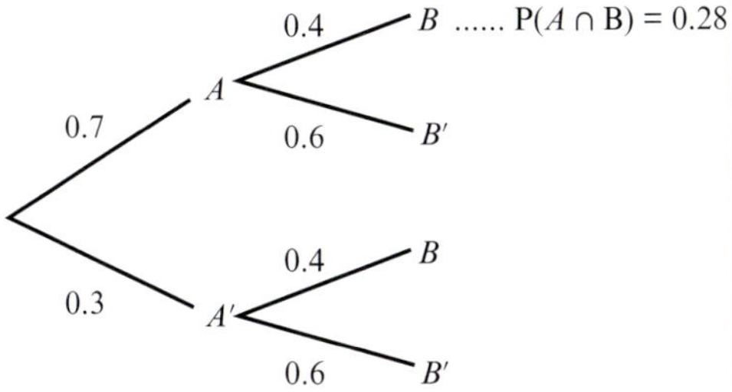

c 0.18

4 $\frac{5}{12}$

5 $\frac{3}{8}$

6 **a** $\frac{1}{16}$ **b** $\frac{15}{16}$ **c** $\frac{671}{1296}$

7 $\frac{35}{66}$

8 **a** $\frac{7}{216}$ **b** $\frac{103}{108}$

Exercise 4D

1 0.084

2 0.081

3 0.75

4 **a** 0.06 **b** 0.56

5 **a** Proof

b $P(A) = \frac{7}{16}$, $P(B) = \frac{11}{16}$ and $\frac{7}{16} \times \frac{11}{16} \neq \frac{1}{8}$

6 **a** $\frac{16}{69}$

b $\frac{44}{69} \times \frac{27}{69} \neq \frac{16}{69}$

c $n = 3; \frac{44}{66} \times \frac{24}{66} = \frac{16}{66}$

7 **a** The data do not support the theory; $\frac{7}{17} \times \frac{7}{17} \neq \frac{3}{17}$

b A person's ability to dance salsa depends on whether or not they wear spectacles.

c This is unlikely to be true. Different proportions of people (42.9% and 40%) can dance salsa due to chance, not because they do or do not wear spectacles.

8 **a** $x = 12$

b Proof, using probabilities as in the following table:

	S	*S′*	Totals
B	12	28	40
B′	18	42	60
Totals	30	70	100

c Proof, using probabilities as in the following table:

	S	*S′*	Totals
B	x	$b - x$	b
B′	$s - x$	$T - b - s + x$	$T - b$
Totals	s	$T - s$	T

Exercise 4E

1 **a** $\frac{4}{7}$ **b** $\frac{1}{3}$

2 $\frac{1}{3}$

3 **a** $\frac{5}{8}$ **b** $\frac{1}{2}$

4 $\frac{2}{3}$ **5** $\frac{6}{7}$

6 **a** $\frac{25}{28}$ **b** $\frac{29}{35}$ **c** $\frac{29}{70}$

7 $\frac{5}{6}$ **8** $\frac{9}{13}$

Exercise 4F

1 **a**

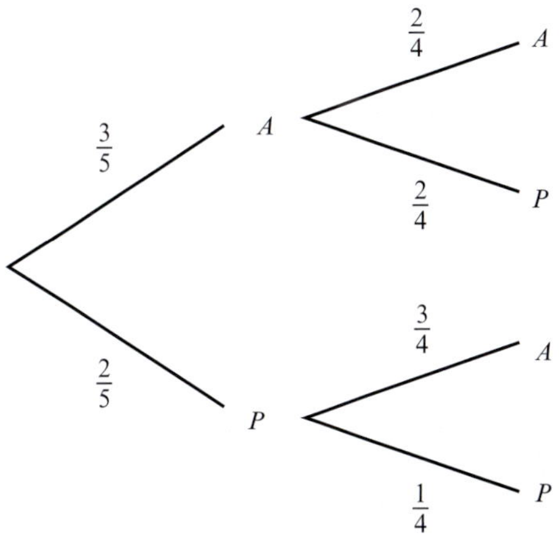

b $\frac{3}{4}$ **c** $\frac{2}{5}$

2 **a** B = black sock, R = red sock

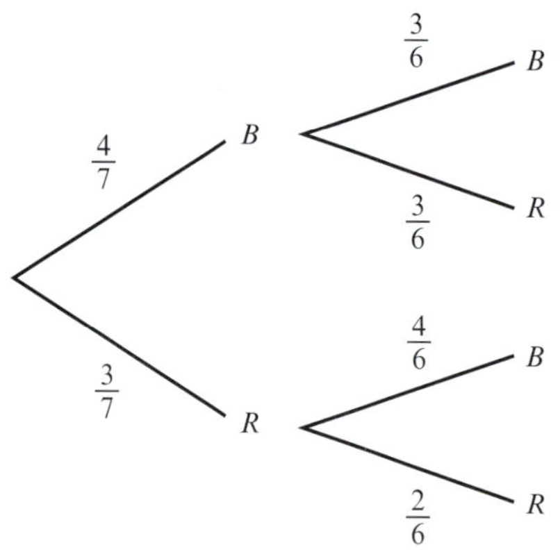

b **i** $\frac{2}{7}$ **ii** $\frac{1}{7}$ **iii** $\frac{4}{7}$

3 **a** 0.35 **b** 0.41 **c** 0.94

4 $\frac{7}{17}$

5 $\frac{13}{35}$

6 **a** 0 **b** $\frac{1}{9}$ **c** $\frac{1}{18}$ **d** $\frac{5}{12}$

7 **a** 0.21762 **b** 0.10934

8 **a** $\frac{1}{5}$ **b** $\frac{1}{9}$

End-of-chapter review exercise 4

1 $\frac{2}{3}$

2 $\frac{5}{7}$

3 **a** **i** $\frac{3}{10}$ **ii** $\frac{9}{20}$ **iii** $\frac{5}{16}$

b $\frac{7}{22}$

4 **a** **i** $\frac{1}{8}$ **ii** $\frac{1}{4}$

b $\frac{5}{18}$

5 **a** $\frac{22}{1365}$ or 0.0161 **b** $\frac{19}{65}$ or 0.292

c $\frac{88}{4095}$ or 0.0215

6 **a** **i** 0.18 **ii** 0.3439

b 0.0022

7 **a** $\frac{5}{13}$ **b** $\frac{247}{459}$ or 0.538

c Either: P = C, Q = A, X = Chemistry, Y = Biology
Or: P = A, Q = C, X = Biology, Y = Chemistry

8 $\frac{3}{8}$

9 **a** $\frac{27}{32}$ **b** $\frac{13}{18}$

10 $\frac{138}{400} \neq \frac{175}{400} \times \frac{322}{400}$ to show that P(W and CC) $\neq$ P(W) $\times$ P(CC)

5 Permutations and combinations

Exercise 5A

1 120

2 2

3 **a** 432 **b** $\frac{5}{18}$

4 $\frac{8!}{6!} + \frac{6!}{4!}$

5 19 cm

6 $(2!)^3 \times (4!)^2$ or $(2!)^7 \times (3!)^2$

7 Proof

8 $\theta = 60°, 120°, 240°$ or $300°$

9 $a = 3, b = 5$

Exercise 5B

1 5040

2 720

3 **a** 720 **b** 24 **c** 3 628 800

4 **a** 4.03×10^{26} **b** 479 001 600

c 2.04×10^{46}

5 2.65×10^{32}

6 30

7 24

8 They have not allowed for the fact that the sculptures do not have to face in the same direction.
$(3 \times 4) \times (2 \times 4) \times (1 \times 4)$ or $3! \times 4^3 = 384$

Exercise 5C

1 **a** 83 160 **b** 3.06×10^{22}

2 113 400

3 The four canaries and the three love-birds are seven distinct objects.

4 362 880

5 a 1296 b 81

6 a 15 b 375 c 109 296 000

7 a They are front and rear views of the same arrangement.

b 20

8 a 958 003 200

b i 4 ii 6.80×10^{44}

Exercise 5D

1 a 24

b i 18 ii 12 iii 10

2 40 320

3 7 257 600

4 a 420 b 2

5 a 8.72×10^{10} b 958 003 200

c 435 456 000

6 20 160

7 a 3 628 800 b 402 796 800

c 79 833 600

8 a 2520 b 5040

c 60 480

9 a 20 159 b 12

Exercise 5E

1 a 1716 b 6.08×10^{16}

2 a 360 b 2160

c 1800 d 720

3 11 880

4 a 6 b 3, 5 and 7

5 a 4 b 60

6 a 9 b 30

7 36

8 a 1 814 400 b 282 240

9 73

Exercise 5F

1 a 71 b 840 c 2002

2 a 35 b 1

3 a 120 b 70 c 8400

4 a 167 960 b 252 c 155 040

5 a 462

b i 200 ii 281

6 a 1287 b 3432 c 5148

7 a 45 b 145

8 a 1.48×10^{12} b 11 027 016

c 7 332 965 640 or 7.33×10^9

9 $n = r + 1$

Exercise 5G

1 0.999

2 a 10 b 6 c 0.9

3 a $\frac{133}{299}$ or 0.445 b $\frac{1}{2}$

4 a $\frac{n}{2}(n-1)$ b 78

5 a 1.19×10^{24} b 6912

6 789

7 a 0.0112 b 0.342

8 $\frac{2}{21}$

9 a Proof

b

n	0	1	2	3	4	5
${}_n\mathrm{P}(A)$	1	0	$\frac{1}{4}$	0	$\frac{9}{64}$	0

${}_n\mathrm{P}(A) = 0$ for all odd n

c

n	0	2	4	6	8	10
${}_n\mathrm{P}(A)$	1	$\frac{1}{4}$	$\frac{9}{64}$	$\frac{400}{4^6} = \frac{25}{256}$	$\frac{4900}{4^8} = \frac{1225}{16384}$	$\frac{63504}{4^{10}} = \frac{3969}{65536}$

d ${}_{n+2}\mathrm{P}(A) = \left(\frac{n+1}{n+2}\right)^2 \times {}_n\mathrm{P}(A)$

End-of-chapter review exercise 5

1 a 20 b 25

2 a 10 080 b 9360

3 99 768 240

4 a 240 b 144

5 7

6 a 243 b $\frac{2}{27}$

7 17

8 a 440 b $\frac{10}{11}$

9 $\frac{3}{13}$

6 Probability distributions

Exercise 6A

1 $k = 0.3$

2 a 0.33 b 0.19

3 a Proof b $k = 0.12$, $k = 0.06$

c $k = 0.06$ is valid; $k = 0.12$ gives $\mathrm{P}(V = 4) < 0$.

d 0.92

4 $k = \frac{24}{25}$, $T = 2$

5 a Proof b $\frac{21}{22}$

6 a $a = 0.64$, $b = 0.82$, $c = 0.94$
b 0.36

7 a 0.7
b

a	0	1	2
$P(A = a)$	0.36	0.48	0.16

8 a Proof
b

x	0	1	2	3
$P(X = x)$	$\frac{8}{125}$	$\frac{36}{125}$	$\frac{54}{125}$	$\frac{27}{125}$

c 2

9 $k = \frac{3}{14}$ is invalid because it gives $P(T = 5) < 0$.

Exercise 6B

1 4.5

2 a $a = 9$, $b = 10$
b i

t	3	7	11	15	19
$P(T = t)$	$\frac{1}{9}$	$\frac{2}{9}$	$\frac{3}{9}$	$\frac{2}{9}$	$\frac{1}{9}$

ii $E(T) = 11$, $Var(T) = 21\frac{1}{3}$
iii 0

3 a Proof
b i

c	0	1	2
$P(C = c)$	$\frac{6}{15}$	$\frac{8}{15}$	$\frac{1}{15}$

ii Proof; $E(C) + E(P) = 2$

4 $a = 43$, $Var(R) = 77.76$

5 $E(Y) = m + 1.1$, $Var(Y) = 0.49$ for all values of m

6 Exp = \$28 000; SD = \$33 900. Quite likely to generate over \$25 000, but it is a high-risk venture.

7 a $\left(\frac{1}{4} \times \frac{10}{36}\right) + \left(\frac{1}{2} \times \frac{2}{6}\right)$
b $\frac{19}{36}$
c

Score	0	1	2	3	4	5
Probability	$\frac{7}{24}$	$\frac{17}{72}$	$\frac{5}{36}$	$\frac{5}{24}$	$\frac{1}{36}$	$\frac{7}{72}$

SD = 1.59

8 a

h	0	1	2
$P(H = h)$	$\frac{10}{66}$	$\frac{35}{66}$	$\frac{21}{66}$

k	0	1	2	3
$P(K = k)$	$\frac{2}{44}$	$\frac{14}{44}$	$\frac{21}{44}$	$\frac{7}{44}$

b Kara; $E(H) = 1\frac{1}{6}$ and $E(K) = 1\frac{3}{4}$

9 a

s	0	1	2
$P(S = s)$	0.36	0.48	0.16

h	0	1	2
$P(H = h)$	0.09	0.42	0.49

b $(0.36 \times 0.42) + (0.48 \times 0.09) = 0.1944$
c

w	0	1	2	3	4
$P(W = w)$	0.0324	0.1944	0.3924	0.3024	0.0784

d $E(S) = 0.8$, $E(H) = 1.4$, $E(W) = 2.2$
e $E(W) = (2 \times 0.4) + (2 \times 0.7)$

10 a

b	0	1	2
$P(B = b)$	$\frac{2}{15}$	$\frac{8}{15}$	$\frac{5}{15}$

b i $p = 12$, $q = 30$
ii Proof
iii $k = 11$

11 a $a = 8$
b $k = 1001$; \$15.43
c 0.6

12 Any suggestion that knowledge of expectation (≈ 11) and SD(≈ 4) may improve efficiency and levels of care (availability of staff, equipment, medicines, transport, beds etc.)

13 a $b = 4000$, $k = 62$
b

v	14	22	34	54	62
$P(V = v)$	$\frac{951}{2851}$	$\frac{879}{2851}$	$\frac{711}{2851}$	$\frac{271}{2851}$	$\frac{39}{2851}$

$SD(V) = 12.7$, $\frac{1}{2}\overline{V} = 13.0$

End-of-chapter review exercise 6

1 a $\frac{9}{20}$ or 0.45

b $(0 \times 0.1) + (1 \times 0.2) + (2 \times 0.3) + (3 \times 0.4) = 2$

c 1

2 a $\frac{1}{15}$

b Mean $= 1\frac{1}{3}$, SD = 1.25

3 a $k = \frac{1}{4a + 12}$ b 8.625

4 a $2p$

b $2p(1 - p) + 4p^2 - (2p)^2 = 2p(1 - p)$

5 a

x	0	1	2
$P(X = x)$	$\frac{3}{21}$	$\frac{12}{21}$	$\frac{6}{21}$

b $1\frac{1}{7}$ and $\frac{20}{49}$

c $E(Y) = \frac{6}{7}$, $Var(Y) = \frac{20}{49}$

6 a $a + b = 0.77$

$a + 2b = 1.18$

$a = 0.36, b = 0.41$

b 0.40

7 a $\left(\frac{3}{7} \times \frac{5}{13} \times \frac{4}{12}\right) + \left(\frac{4}{7} \times \frac{6}{13} \times \frac{5}{12}\right) = \frac{15}{91}$

b

f	0	1	2
$P(F = f)$	$\frac{15}{91}$	$\frac{48}{91}$	$\frac{28}{91}$

c 0.452

8 0.114

7 The binomial and geometric distributions

Exercise 7A

1 a 0.269 b 0.731

2 a $\frac{864}{2401}$ b $\frac{513}{2401}$

3 a 0.216 b 0.340 c 0.611

4 $P(X = 1) = \frac{n}{2^n}$

$P(X = 2) = \frac{n(n - 1)}{2^{n+1}}$

5 a $P(X = 1) = 0.154$, $P(Y = 1) = 0.432$

b X and Y are independent.

c 0.148

6 0.653

7 0.797

8 12

9 0.2745

10 a Male and female lambs are equally likely. Sexes of different lambs are independent of each other.

b 0.0547

11 a No. The probability that a train arrives late depends on, for example, whether it departed on time.

b Each train's arrival time is independent of all other trains' arrival times. This may not be a reasonable assumption; for example, one train arriving late may cause another train's arrival to be delayed.

12 a Proof b $n = 13$; 0.246

Exercise 7B

1 a $E(V) = 1.8$, $Var(V) = 1.26$

b $E(W) = 6.3$, $Var(W) = 3.654$

c $E(X) = 153$, $Var(X) = 22.95$

d $E(Y) = 32.12$, $Var(Y) = 8.6724$

2 a $E(X) = 8$, $Var(X) = 7$

b 0.149

3 a 0.885 b 0.577

4 a $n = 60, p = 0.45$ b 0.0759

5 a $n = 84, p = \frac{7}{12}$ b 0.0863

6 $n = 3, p = \frac{5}{6}$

v	0	1	2	3
$P(V = v)$	$\frac{1}{216}$	$\frac{15}{216}$	$\frac{75}{216}$	$\frac{125}{216}$

7 a 16 b 0.120

8 a Proof

b The biologist ignored the fact that 45 samples are already known to contain the microorganism. B(400, 0.045) should be used to find $P(X = 18)$, rather than B(1400, 0.045) to find $P(X = 63)$. Probability is 0.0958.

9 a 0.238 is an underestimate

b 0

Exercise 7C

1 **a** 0.026568 **b** 0.973432 **c** 0.9676

2 **a** 0.00567 **b** 0.973

3 **a** $p = 0.5$ **b** 0.875

4 **a** 0.64 or $\frac{16}{25}$

b 0.136125 or $\frac{1089}{8000}$

5 **a** No; 450 is significantly greater than the expected number of 6s (200).

b $\frac{3}{8}$ or 0.375

c **i** 0.0891 **ii** 0.146

6 **a** 15 **b** Proof **c** 0.105

7 **a** 0.128 **b** 0.008

8 **a** 0.5184 **b** 0.7284 **c** 0.79456

9 $\frac{177}{320}$ or 0.553

Exercise 7D

1 0.625

2 **a** 0.2 or 0.8 **b** 5 or 1.25

3 0.185

4 0.0949

5 **a** Proof **b** One **c** $\frac{5}{9}$

6 **a** \$78.75 **b** 0.0864

7 **a** **i** Once **ii** $10\frac{2}{3}$ times

b 13

8 **a** 8 **b** 0.118

9 $a = \frac{3}{2}, b = -\frac{3}{4}$

End-of-chapter review exercise 7

1 **a** 0.233 **b** 0.355

2 **a** $E(D) = 2.55$, $Var(D) = 0.3825$

b 0.325125

3 $(1 - \sqrt{0.6724})^2 = 0.0324$ or $2 \times \sqrt{0.6724} \times \sqrt{0.0324} = 0.2952$

4 **a** Mode = 1, expectation = $5\frac{1}{2}$

b $\frac{162}{1331}$ or 0.122 **c** 0.201

5 $a = 4$

6 0.573

7 **a** That they come to feed independently and at random.

b 0.523 **c** 0.0752

8 **a** (2, 1), (1, 2) and (0, 3)

b 0.749

9 **a** 0.3 **b** $\frac{11}{84}$ or 0.131

c $\frac{803}{7056}$ or 0.114 **d** 0.379

8 The normal distribution

Exercise 8A

1 **a** Continuous **b** Discrete

c Neither **d** Continuous

2 **a** **i** $\mu = 15$ **ii** $\sigma > 3$

b **i** $P(X > 19)$ **ii** $P(11 < Y < 19)$

3 **a** $P(T < 56) = P(T > 56)$

b $P(T > 60) > P(T < 50)$

c $P(20 < T < 92) \approx 1$

d $P(47 < T < 51) < P(58 < T < 62)$

4 Same width and height; B centred 4 units to the right of A.

5

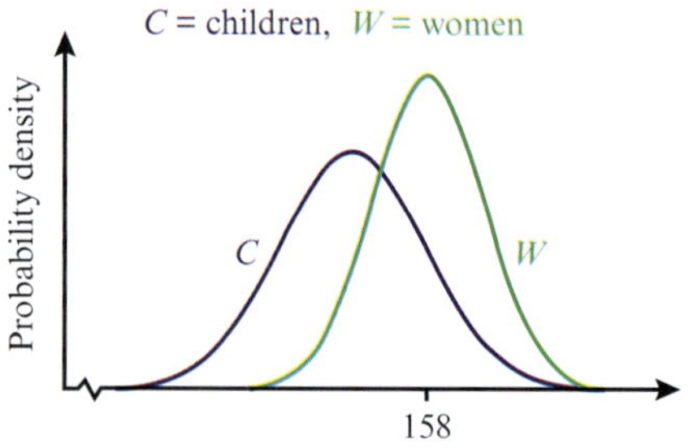

6 The mean decreased to 12.65, but the variation did not change.

7 **a** **i** $\mu > m$ **ii** $\sigma < s$

b Y is not normally distributed.

8 $Y \sim N(0.382, 0.00006)$; the shoes in each pair have identical masses.

Exercise 8B

1 **a** 0.891 **b** 0.993 **c** 0.562

d 0.108 **e** 0.0087 **f** 0.278

g 0.966 **h** 0.980 **i** 0.527

j 0.0336 **k** 0.003 **l** 0.418

m 0.950 **n** 0.05 **o** 0.950

p 0.050

2 **a** 0.0366 **b** 0.120 **c** 0.342

d 0.439 **e** 0.956 **f** 0.780

3 **a** 0.856 **b** 0.209 **c** 0.0320

d 0.119 **e** 0.446 **f** 0.950

g 0.980 **h** 0.806 **i** 0.0164

4 **a** 0.440 **b** 1.165 **c** 2.150
d 1.017 **e** 0.240 **f** 1.178
g 2.450 **h** 0.758
i −2.83 or −2.84
j −1.955 **k** −1.035 **l** 0
m −2.74 **n** −2.192 **o** −1.677
p −0.056 **q** 1.645 **r** 1.282
s 2.576 **t** 0.674

5 41500

6 24.2%

7 $k = 1.96$

8 ≈ 1697

9 $k = 0.209$

Exercise 8C

1 **a** $X \sim N(20, 4^2)$ or $N(20, 16)$
b 1.5
c **i** 0.933 **ii** 0.0668

2 **a** 0.940 **b** 0.5 **c** 0.758
d 0.670 **e** 0.119

3 **a** 0.933 **b** 0.0062 **c** 0.773
d 0.0401

4 **a** 0.952 **b** 0.0098 **c** 0.748
d 0.0038

5 **a** 0.136 **b** 0.0606 **c** 0.733
d 0.770 **e** 0.866

6 **a** 0.159 **b** 0.0228

7 **a** $s = 41.6$ **b** $t = 29.9$
c $u = 37.3$ **d** $v = 31.7$

8 **a** Proof **b** $15.05 - \mu = 2.51\sigma$
c $\mu = 2.5, \sigma = 5$ **d** 0.136

9 65.0

10 Mean = 42.0, SD = 8.47

11 $\mu = 9.51, \sigma = 0.303$ **12** 0.954

Exercise 8D

1 **a** 0.0370 **b** 0.0062 **c** 0.821

2 **a** 0.614 **b** $h = 20.9, k = 16.3$
c 17

3 47 hours^2

4 Mean = 49.1, SD = 13.4

5 **a** 0.773 **b** 0.598

6 0.358

7 0.288

8 **a** 82.49 to 91.41cm **b** 0.944

9 **a** 4.60 **b** 0.150

10 0.504

Exercise 8E

1 **a** 62 **b** 23.56
c **i** 70 incorrect; no continuity correction; use 69.5.
ii 0.939

2 **a** $np = 37.4$, $nq = 46.6$
b It can be well-approximated by a normal distribution, i.e. N(37.4, 20.944).
c 0.737

3 **a** 0.0288 **b** 0.114 **c** 0.0805 **d** 0.375
$np = 12$, $nq = 28$, valid normal approximation

4 **a** 0.0083 **b** 0.911
$np = 6$, $nq = 44$, valid normal approximation

5 **a** 0.0168 **b** 33

6 **a** 0.189; binomial; $np = 1$
b 0.642; binomial; $np = 1$
c 0.223; normal; $np = 32.1$, $nq = 17.9$

7 **a** 0.960
b **i** Binomial; $n = 3$, $p = 0.9599$
ii 0.111

8 **a** The number of points scored is a discrete quantity.
b $X \sim B(10, 0.065)$ **c** 0.355

9 **a** **i** 0.108 **ii** 0.7694 **iii** 0.2568
b Most reliable is **i** (not an approximation). Least reliable is **iii** ($p = 0.12$ is furthest from 0.5)

10 **a** 0.125
b Company T; increase the circumference of the tubes or use stretchable material.

End-of-chapter review exercise 8

1 **a** 0.949 **b** 0.0278

2 0.2257

3 **a** 10 **b** 0.791

4 **a** $120 - \mu = 0.842\sigma$
$115 - \mu = 0.063\sigma$
b $\mu = 114.6, \sigma = 6.42$ **c** 0.989

5 **a** $\mu = 53.3, \sigma = 15.2$ **b** 0.330

6 **a** 80.7% **b** 0.211

7 0.0554

8 0.308

9 0.650

10 Proof